THE WORLD
ACCORDING TO
STAR WARS

THE WORLD
ACCORDING TO

CASS R. SUNSTEIN

DEY ST.
AN IMPRINT OF WILLIAM MORROW *PUBLISHERS*

THE WORLD ACCORDING TO STAR WARS. Copyright © 2016 by Cass R. Sunstein. All rights reserved. Printed in the United States of America. No part of this book may be used or reproduced in any manner whatsoever without written permission except in the case of brief quotations embodied in critical articles and reviews. For information, address HarperCollins Publishers, 195 Broadway, New York, NY 10007.

HarperCollins books may be purchased for educational, business, or sales promotional use. For information, please e-mail the Special Markets Department at SPsales@harpercollins.com.

FIRST EDITION

Designed by Shannon Nicole Plunkett

Library of Congress Cataloging-in-Publication Data has been applied for.

ISBN 978-0-06-248422-2

16 17 18 19 20   OV/RRD   10 9 8 7 6 5 4 3 2 1

For Declan—my son

Difficult to see. Always in motion is the future.

—YODA

*It's the biggest adventure you can have,
making up your own life, and it's true for everybody.
It's infinite possibility.*

—LAWRENCE KASDAN

CONTENTS

PREFACE   *XI*

INTRODUCTION: LEARNING FROM STAR WARS   *1*

EPISODE I: I AM YOUR FATHER:
The Heroic Journey of George Lucas   9

EPISODE II: THE MOVIE NO ONE LIKED:
An Expected Flop Becomes the Defining Work of Our Time   31

EPISODE III: SECRETS OF SUCCESS:
Was Star Wars Awesome, Well-Timed, or Just Very Lucky?   39

EPISODE IV: THIRTEEN WAYS OF LOOKING AT STAR WARS:
Of Christianity, Oedipus, Politics, Economics, and Darth Jar Jar   65

EPISODE V: FATHERS AND SONS:
You Can Be Redeemed, Especially If Your Kid Really Likes You   91

EPISODE VI: FREEDOM OF CHOICE:
It's Not About Destiny or Prophecy   103

EPISODE VII: REBELS:
Why Empires Fall, Why Resistance Fighters (and Terrorists) Rise   113

EPISODE VIII: CONSTITUTIONAL EPISODES:
Free Speech, Sex Equality, and Same-Sex Marriage as Episodes   145

EPISODE IX: THE FORCE AND THE MONOMYTH:
Of Magic, God, and Humanity's Very Favorite Tale   161

EPISODE X: OUR MYTH, OURSELVES:
Why Star Wars Gets to Us   177

BIBLIOGRAPHICAL NOTE   *181*
ACKNOWLEDGMENTS   *183*
NOTES   *187*
INDEX   *207*

PREFACE

The human race can be divided into three kinds of people: those who love Star Wars, those who like Star Wars, and those who neither love nor like Star Wars. I have read parts of this book to my wife, emphasizing those that seemed to me especially fun, and one night she finally responded, with some combination of pity and exasperation, "Cass, I just don't love Star Wars!" (I knew that, I guess, but somehow I forgot.)

When I started this book, I merely liked Star Wars. I have now gone way past the threshold for love. Nonetheless, I intend this book for all three kinds of people.

If you love it, and are sure that Han shot first, and know everything there is to know about parsecs, Biggs, Boba Fett, and General Hux, you still might want to learn about the series' unlikely origins, its wildly unanticipated success, and what it really has to say about fathers, freedom, and redemption. If you merely like the movies, you might be interested to know their distinctive claims about destiny, heroic journeys, and making the right choice when the chips are down.

If you really don't like Star Wars, and can't tell an Ackbar from a Finn or a Windu, you might still be curious about how it became such a cultural phenomenon, and about why it's so resonant, and why its appeal is enduring, and how it illuminates childhood, the complicated relationship between good and evil, rebellions, political change, and constitutional law.

In his wild fever-dream "Auguries of Innocence," William Blake wrote of seeing "a World in a Grain of Sand." Star Wars is a grain of sand; it contains a whole world.

THE WORLD
ACCORDING TO
STAR WARS

LEARNING FROM STAR WARS

All the gods, all the heavens, all the hells, are within you.

—JOSEPH CAMPBELL

As of early 2016, the Star Wars franchise had earned about $30.2 billion. Of that amount, $6.25 billion came from box office, nearly $2 billion from books, and about $12 billion from toys. The total exceeds the gross domestic product of about ninety of the world's nations, including Iceland, Jamaica, Armenia, Laos, and Guyana. Suppose that Star Wars was a nation and that its earnings were its GDP. If so, it would be ranked right around the middle of the 193 nations on the planet. Doesn't it deserve its own seat at the United Nations?

Moreover, its earnings are rapidly climbing. With the spectacular success of *The Force Awakens*, they're exploding.

The numbers do not come close to telling the tale. "Quantify everything you cannot." (Didn't Yoda say that?) In terms of politics and culture, Star Wars is everywhere. In the 1980s, Ronald Reagan's Strategic Defense Initiative was commonly called "Star Wars." After the appearance of *The Force Awakens* in December 2015, President Barack Obama closed a news conference this way: "Okay, everybody, I gotta get to Star Wars." In the same week, Hillary Clinton ended the national Democratic debate with the words, "May the Force be with you."

Also in that week, Republican candidate Ted Cruz tweeted, "The Force . . . it's calling to you. Just let it in and tune in to tonight's #CNNDebate." Drawing on westerns and 1960s comic books, committed to liberty, and audacious about hope, Star Wars is bipartisan and all-American.

At the same time, its appeal is universal; obsession with the series is hardly limited to the United States. In late 2015, I traveled to Taiwan to give a series of lectures and to meet with the nation's president and its Constitutional Court. We discussed human rights, regulation, the world economy, and Taiwan's complex relationship with China. But everyone also wanted to talk about Star Wars. The saga is big in France, Germany, Italy, Nigeria, and the United Kingdom. They love it in Israel, Egypt, and Japan, and it's successfully invaded India, too. Until 2015, it wasn't allowed to be shown in China, but the Force has now awakened there as well.

In all of human history, there's never been a phenomenon like Star Wars. Fueled by social media, the whole series has a cultlike following, except that the cult is so large that it transcends the term. It's humanity, just about. A recent Google search of "Star Wars" produced 728 million results. By contrast, "Beatles" produced 107 million, "Shakespeare" 119 million, "Abraham Lincoln" 69 million, "Steve Jobs" 323 million, and "Taylor Swift" 232 million. As evidence of its adaptability, consider the first entry turned up by a Twitter search: "Destroy hunger with Star Wars Death Star peanut butter cup."

Okay, maybe you don't love Star Wars, or even like it. Whether or not you consider yourself a fan, you probably know a lot about the movies. You know about the Force, don't you? You've heard of Darth Vader? Might you confess that in the toughest times, you've sometimes thought to yourself, in your heart of hearts, "Help me, Obi-Wan Kenobi, you're my only hope"?

Star Wars unifies people. You can live in Berlin or New York, London or San Francisco, Seattle or Paris, and you'll probably recognize Darth Vader, and you might well know what the Millennium Falcon is. (You do, don't you?) In 2015, the United States and Russia were not especially friendly; there was a great deal of tension between Vladimir Putin and Barack Obama. But when *The Force Awakens* came out, a high-level Russian official told me, with a bright, boyish smile and something like an acknowledgment of common humanity, that the series is truly beloved in his country, and that just about everyone there has seen it.

Star Wars also connects generations. My three-year-old daughter, named Rian, really likes Darth Vader. My six-year-old son, named Declan, loves to wield his lightsaber. I also have a grown daughter, named Ellyn, with whom I saw the first two trilogies—starting, I think, when she was around seven years old. She texted me these words, right after seeing *The Force Awakens:* "I burst into tears right in the opening credits. . . . First time we didn't go together!"

My own parents are long gone, but my mother, a science fiction buff, adored the first movie, *Star Wars*, released in 1977 (and later retitled *A New Hope*). It seemed a little more confusing to my father, a naval officer in World War II, stationed in the Pacific in the 1940s, who knew how to handle a gun (he fought in the Clone Wars, kind of). My dad loved tennis, cars, and fishing—not so much lightsabers and droids. But he was game for anything, and he saw the movie's charm. Healthy as a Wookie, he was stricken with brain cancer in his early sixties, and he died young, just four years after *A New Hope* was released. He never had a chance to meet Ellyn, Declan, or Rian.

Different cultures have different rituals and traditions. There's Santa Claus, of course, and the Easter Bunny, and the Tooth Fairy. But nothing quite compares to sitting with a young child for her first viewing of Star Wars. As the lights come down, and those beloved golden letters fill the screen, and John Williams's familiar music announces what's coming, there's awe and wonder. Plenty of ghosts enter the room. It's good to see them. Star Wars brings back the dead.

STAR WARS IS OUR MODERN MYTH

When *A New Hope* was released, most insiders thought that it would be a disaster. The studio had no faith in it. Almost nobody liked it. The actors thought that it was ridiculous. George Lucas, its creator, feared catastrophe. All this raises some questions: Why has Star Wars turned out to be such a spectacular success? Is it really awesome? Why is its appeal so enduring? Why has it become our modern myth? What does it have to teach us? About culture, psychology, freedom, history, economics, rebellion, human behavior, and law? About the human heart?

I'll try to answer all of these questions. It helps that like a poem or a novel, the Star Wars saga leaves plenty of room for diverse interpretations. Is it a critique of empires and a fervent plea for democracy, or just the opposite? Is it really in favor of the Light Side, or is it secretly in love with the Dark? Is the Force God, or is it something inside each of us? What do the movies have to say about Christianity? About gender and race? About capitalism? About the meaning of loyalty? About why history takes the turns that it does?

Star Wars offers a modern version of a universal tale: the Hero's Journey. Lucas was self-conscious about this, drawing directly on Joseph Campbell's wildly influential book, *The Hero with a Thousand Faces*, which sets out the central life events that unify countless myths. (Lucas described Campbell as "my Yoda.") In its essentials, the hero's journey is the tale of Jesus Christ, Buddha, Krishna, and Muhammad—and also Spider-Man, Superman, Batman, Jessica Jones, and Luke Skywalker (and Anakin, too, and also Rey, and possibly Finn and Kylo).

The Hero's Journey has deep psychological resonance. It taps directly into the recesses of the human psyche. Whoever you are, it's your tale as well. (You'll see.)

At the same time, Star Wars is keenly alert to the immense power of the two sides of the Force. It demonstrates that for all of us, the choice between Light and Dark is far from simple. (You're kidding yourself, and not living a full life, if you think that it is. Every human being needs to visit the Dark Side. Try it. Don't linger.) Stylized though it is, the series has important things to say about republics, empires, and rebellions. It knows that republics can be highly fragile, and empires, too—and that the success of rebellions often turns on small decisions and seemingly irrelevant factors.

Star Wars is obsessed with the complex relationship between fathers and sons, and with what they will do for each other, especially when life itself hangs in the balance. On that count, its lessons are powerful and enduring. Before long, it's going to say something powerful about daughters as well. (After *The Force Awakens*, it's almost there.) When parents see the saga with their children, young or old, they are having a blast, but they are also learning, and feeling, something important about the nature of their attachment.

FREE TO CHOOSE

Star Wars also makes a bold claim about freedom of choice. Whenever people find themselves in trouble, or at some kind of crossroads, the series proclaims: *You are free to choose.* That's the deepest lesson of Star Wars. That's the twist on the Hero's Journey. The emphasis on freedom of choice, even when things seem darkest and life is most constrained, is the saga's most inspiring feature. It also has everything to do with the theme, central to the saga, of forgiveness and redemption. (According to Star Wars: you can always be forgiven, and you can always be redeemed.)

The great scriptwriter Lawrence Kasdan, Lucas's collaborator on *The Empire Strikes Back* and *Return of the Jedi,* as well as J. J. Abrams's collaborator on *The Force Awakens,* puts it this way, with childlike wonder: "It's the biggest adventure you can have, making up your own life, and it's true for everybody. It's infinite possibility. It's like, I don't know what I'm going to do in the next five minutes, but I feel I can get through it. It's an assertion of a life force."

A lot of people think that the Star Wars movies are about destiny and the ultimate authority of prophecy. Not at all. "Impossible to see, the future is." (Yoda actually says that.) That's the hidden message and the real magic of Star Wars—and the foundation of its rousing tribute to human freedom.

THE PLAN

I'm going to be covering some diverse topics here, including the nature of human attachment, whether timing is everything, how to rank the seven Star Wars movies, why Martin Luther King Jr. was a conservative, how boys need their mothers, the workings of the creative imagination, the fall of Communism, the Arab Spring, changing understandings of human rights, whether *The Force Awakens* was a triumph or a disappointment, the limits of human attention, and whether Star Wars really is better than Star Trek.

For those who like roadmaps: Episodes I, II, and III explore how George Lucas came up with the Star Wars saga and why *A New Hope* turned out, against all odds, to be such a smashing success. Episodes IV, V, and VI examine the films' intriguingly multiple meanings and also what they have to say about their three most important topics: fatherhood, redemption, and freedom. Episodes VII and VIII turn to the saga's lessons for politics, rebellions, republics, empires, and constitutional law. Episodes IX and X investigate magic, behavioral science, and the Force—and why Star Wars turns out to be timeless.

I AM YOUR FATHER

The Heroic Journey of George Lucas

Two roads diverged in a yellow wood,
And sorry I could not travel both
And be one traveler, long I stood
And looked down one as far as I could
To where it bent in the undergrowth. . . .

—ROBERT FROST

People often think that for important texts, there is some kind of grand designer, who figured it all out in advance, and whose essential plan is responsible for everything that follows. Maybe the designer is a person: William Shakespeare, Leonardo da Vinci, Jane Austen, George Washington, Steve Jobs, or J. K. Rowling. Or maybe the designer is an institution: Wall Street, Congress, the market, the CIA, or Hollywood.

The truth is that the best designers tend to be improvisers. They have ideas, and they shoot off sparks, but they may have nothing that counts as a grand plan. Like Luke, Han, Anakin, and Rey, they make choices on the spot. They start things, which end up going in all sorts of unanticipated directions. Characters and plots seem to create their own momentum, even to tell their own tales.

Designers might have a sense of ultimate destination, a kind of internal GPS, or a closing image, but the work itself might take them way off their expected course—and they will be immensely grateful for that. As much as anyone, they might be surprised, even astonished, by the shape that their design ends up taking. They encounter forks in the road, and they opt for one path rather than the other. That's how creativity works.

Behavioral scientists refer to the "planning fallacy," which means that people typically think that a project will get finished a lot faster than it actually does. In the words of the great psychologists Daniel Kahneman and Amos Tversky, "Scientists and writers, for example, are notoriously prone to underestimate the time required to complete a project, even when they have considerable experience of past failures to live up to planned schedules."

Ask any high school student working on a paper, or any city operating a construction project, or any engineer trying to design a Millennium Falcon, and you're likely to see the planning fallacy in action. But with respect to the creative imagination, there is a different and far more interesting kind of planning fallacy. Call it the *myth of creative foresight*. You end up having to make choices that surprise you; the directions you anticipate turn out not to be the directions in which you go. You can't plan it out in advance.

That's true of the creation of the Star Wars series. It's also a major lesson for both characters and viewers. Farm boy Luke, a Jedi master? Han Solo, no longer a Solo? Darth Vader, redeemed? Finn, helping the Resistance? Scavenger Rey, wielding Luke's lightsaber?

Who would have thought?

"I WANTED TO DO FLASH GORDON"

In explaining how he came up with the Star Wars movies, George Lucas has said different things over the years.

Here's one of his accounts:

> You have to remember that originally, Star Wars was intended to be one movie, Episode IV of a Saturday matinee serial. You never saw what came before or what came after. It was designed to be the tragedy of Darth Vader. It starts with this monster coming through the door, throwing everybody around, then halfway through the movie you realize that the villain of the piece is actually a man and the hero is his son. And so the villain turns into the hero inspired by the son. It was meant to be one movie, but I broke it up because I didn't have the money to do it like that—it would have been five hours long.

Here's another, which is subtly different:

> The Star Wars series started out as a movie that ended up so big that I took out each act and cut it into its own movie. . . . The original concept really related to a father and a son, and twins—a son and a daughter. It was that relationship that was the core of the story. . . . When I first did Star Wars I did it as a big piece.

Yet another, from Lucas's introduction to an edition of the three novelizations of the first trilogy:

> From the outset I conceived Star Wars as a series of six films,
> or two trilogies. . . . When I wrote the original Star Wars
> screenplay, I knew that Darth Vader was Luke Skywalker's
> father; the audience did not. I always felt that this revelation,
> when and if I got the chance to make it, would be startling. . . .

The full story of how Lucas came up with Star Wars is a lot more complicated—and far more interesting. In its earliest versions, Star Wars was not designed to be the tragedy of Darth Vader. Nothing started with a monster coming through the door. There was not a word about a heroic son with a villainous father. Darth Vader, as we know him, came to Lucas's mind relatively late, well after he had the idea for Star Wars—and Vader was a bit player. When Lucas says, "The *Star Wars* story is really the tragedy of Darth Vader," he wasn't lying, but it took him a long time to get there.

The development of the arc of the first trilogy shows Lucas's distinctive combination of obsessiveness, vision, high standards, and dogged willingness to keep learning—along with a form of genius. Lucas never much liked writing; he's a visual person, and dialogue does not come easily to him. Writing the script of *A New Hope* took him several years, and it was a truly miserable experience, almost a form of torture. He closed himself in a room for hours every day, and he forced himself to produce pages, and he hated much of the whole experience. He made himself sick, and he pulled out his own hair (literally). But somehow, this visual artist wrote something iconic.

When Lucas began writing, his thoughts were abstract and vague. In the early 1970s he publicly spoke of his plan for "The Star Wars," in the form of "a western movie set in outer space," or "a Flash Gordon kind of science fiction movie." He described himself as "a big fan of Flash Gordon and a believer in the exploration of space." In 1973, Lucas proclaimed, "*The Star Wars* is a mixture of *Lawrence of Arabia*, the James Bond films, and *2001*. The space aliens are the heroes, and Homo Sapiens naturally the villains."

Except they aren't. At the very beginning, Lucas wanted to purchase the rights to Flash Gordon, to produce a contemporary version—but he couldn't afford them. In his words: "I wanted to do Flash Gordon and tried to buy the rights to it from King Features, but they wanted a lot of money for it, more than I could afford then."

The writing came in fits and starts. The first synopsis was completed in May 1973; the first rough draft appeared a full year later. Neither bore much resemblance to what became *A New Hope*. "I wrote the first version of Star Wars, we discussed it, and I realized I hated the script. I chucked it and started a new one, which I also threw in the trash. That happened four times with four radically different versions," Lucas said. Even after the essential plot of *A New Hope* itself was hatched, the ultimate trajectory of the series, and the Tragedy of Darth Vader, was far from Lucas's mind. By some accounts, *A New Hope* was expected to be a stand-alone movie—*Star Wars*, not Episode IV. Lucas's collaborator Gary Kurtz reveals, "Our plan was to do *Star Wars* and then make *Apocalypse Now* and do a black comedy in the vein of *M*A*S*H*."

Star Wars does turn out to be a tale of fathers and sons, and of a heroic father inspired (and redeemed) by his son, but Lucas thought of those fabulous ideas relatively late—and they changed everything.

XENOS, THORPE, AND THE PRINCE OF BEBERS

At one of the earliest stages of the writing, Lucas produced a list of a large number of (terrific) names, some of which were discarded, including:

Emperor Ford Xerxes the Third

Xenos

Monroe

Mace

Valorum

Biggs

Cleg

Han Solo ("leader of the Hubble people")

Thorpe

Roland

Lars

Kane

Anakin Skywalker ("King of Bebers")

Luke Skywalker ("Prince of Bebers")

There was an ice planet named "Norton III," and a jungle world called "Yavin" (complete with eight-foot-tall Wookies), and a desert planet named "Aquilae."

Early on, Lucas apparently had just one scene clear in his mind, a kind of dogfight in outer space, in which ships "would hurtle and tumble around after each other, like World War II fighters, like wild birds." With the various names in mind, he produced a synopsis, under the heading "Journal of the Whills." The Journal is shrouded in mystery, and there are diverse accounts of its length and its contents.

Apparently it was just a two-page fragment, which began: "This is the story of Mace Windy, a revered Jedi Bendu of Ophuchi who was related to us by C. J. Thape, padawaan learner to the famed Jedi." C. J. stood for "Chuiee Two Thorpe of Kissel," whose father was "Han Dardell Thorpe, chief pilot of the renown galactic cruiser *Tarnack*." (A

Chuiee—but not a Wookie! A Han—a pilot no less, but not a Solo! A Kissel, not a Kessel, and no Kessel Run!) Mace Windy had been "Warlord to the Chairman of the Alliance of Independent Systems. . . . Some felt he was even more powerful than the Imperial Leader of the Galactic Empire. . . . Ironically, it was his own comrades' fear . . . that led to his replacement . . . and expulsion from the royal forces."

In this brief document, Mace and C. J. have their "greatest adventure," which is to act as guardians for "a shipment of fusion portables to Yavin," from which they are "summoned to the desolate second planet of Yoshiro by a mysterious courier from the Chairman of the Alliance." Lucas's initial tale offers little more from there.

It's not exactly clear why this is our heroes' greatest adventure. In fact it sounds pretty terrible. When Lucas began, he didn't have much. But he was clearly inspired. Like most people who are about to produce great things, he seems to have felt a kind of tickle, or maybe an itch. Mace and C. J. weren't destined for greatness, but the itch needed scratching.

BICKERING BUREAUCRATS

Lucas's agent was utterly baffled by the story, so Lucas started on something new, which borrowed heavily from a 1958 movie, *The Hidden Fortress*, by Japan's Akira Kurosawa. His early draft was modeled directly on that movie. Kurosawa's plot was told from the perspective of two bickering peasants, who begin his story wandering a desert landscape during a time of civil war. It's a clever conceit. As Lucas has acknowledged, R2-D2 and C-3PO draw heavily from those characters.

A flavor of Lucas's early text: "The two terrified, bickering bureaucrats crash land on Aquilae while trying to flee the battle of the space fortress." His fourteen-page treatment was so close to *The Hidden Fortress* that it could well be counted a remake, and he considered purchasing the right to the film.

In this version, Lucas did not set "The Star Wars" (as he called it) "a long time ago in a galaxy far, far away." On the contrary, the setting was the distant future: "It is the thirty-third century, a period of civil wars in the galaxy. A rebel princess, with her family, her retainers, and the clan treasure, is being pursued."

The rebel princess travels with a general by a familiar name: Luke Skywalker. Their travels are set in the midst of a battle between an Empire and rebel forces. In one scene, General Skywalker uses a "lazer sword" to kill a bully, who is harassing one of a group of boy rebels at a cantina near a spaceport. (Sounds familiar? A bit?) In another, General Skywalker and the rebels commandeer a squadron of fighter ships. Disguising themselves as Imperial rangers, they end up finding the prison complex of Alderaan, which is the capital of the Empire. Happily, they free the princess, and there is a ceremony at the end, in which the princess stands revealed as a kind of goddess.

It reads a little like Star Wars, but there isn't any R2-D2 or C-3PO, and we don't have most of the other recognizable names, let alone the familiar arc of the plot. In the early stages of the script, Darth Vader was a mere general, relatively unimportant, and blown up at the end. Even at the latest stages of the drafting, Lucas did not think that Darth Vader was Luke's father. Nor did he seek to preserve ambiguity on that count. Actually he thought that Darth Vader was *not* Luke's father. It is a myth, though a good one, that "Darth Vader" is a play on "Dark Father."

A FAIRY TALE

By the way, two robots do appear in some of the early scripts, and they look familiar, and they have quite familiar names: ArTwo Deeto and SeeThreePio. They converse, in a way that's also a bit familiar:

ARTWO: You're a mindless, useless philosopher. . . . Come on! Let's go back to work; the system is all right.
SEETHREEPIO: You overweight glob of grease. Quit following me. Get away. Get away.

Notice anything? ArTwo speaks! And here's something Lucas reported about one of his drafts: "I put this little thing on it: 'A long time ago in a galaxy far, far away, an incredible adventure took place.' Basically, it's a fairy tale now."

In the early drafts of what became *A New Hope*, Darth Vader was not a major character, and Lucas had no act of redemption in mind. That Episode was Luke's tale, not Vader's. The whole idea of redemption came quite late, and it ultimately turned the tale into an improbably moving one about freedom of choice and fathers and sons. Even in the early drafts of *Return of the Jedi*, Vader was not redeemed at all, but was instead rendered irrelevant, as Grand Moff Jerjerrod became the Emperor's new favorite. Nor did Lucas have any sense, at the initial stages, that Palpatine would turn out to be a Sith Lord.

And what about that Journal of the Whills, described by Michael Kaminski as "perhaps the most curious and mysterious item in the history of Star Wars, a thing so shrouded in legend and mystery that it has become a sort of Holy Grail"? Actually it doesn't exist, and it never did, except as that very brief early fragment. But it seemed to ground the whole project, and to give it a kind of gravity, Flash Gordon or no Flash Gordon.

"DON'T TELL ANYONE"

With the "I Am Your Father" moment, Lucas chose to take Star Wars on a fresh narrative path, one that fit well (enough) with what had gone before, but that cast an entirely new light on it, and that was essentially unanticipated by Lucas himself. In an interview after the release of *A New Hope*, Lucas said that he had in mind a sequel "about Ben and Luke's father and Vader when they are young Jedi knights. But Vader kills Luke's father, then Ben and Vader have a confrontation, just like they have in *Star Wars* and Ben almost kills Vader."

Lucas has insisted that he "had Vader in mind as Luke's father all along," but on occasion, he's said something a bit different. Here's a note that he wrote to the writers of *Lost*, the sensational television series: "Don't tell anyone . . . but when 'Star Wars' first came out, I didn't know where it was going either. The trick is to pretend you've planned the whole thing out in advance." Far more revealingly, he acknowledged in 1993, "When you're creating something like that, the characters take over, and they begin to tell the story apart from what you're doing. . . . Then you have to figure out how to put the puzzle back together so it makes sense."

By the way, a lot of writers often speak in just these terms, insisting that their characters "take over" and seem "to tell the story" on their own, with their own integrity and momentum, operating independently of the wishes of the writer. As the lives of their characters develop, they end up on paths that writers could not possibly have foreseen—and hence give the impression of *agency*, even to their authors. William Blake wrote of his works "tho I call them Mine I know they are not Mine," and characterized his process of writing as a kind of dictation, "without Premeditation or even against my Will." Musicians sometimes speak in exactly the same way.

SHIVERS DOWN THE SPINE

How does inspiration work? How does a good story suddenly get deepened, or different? For many creative types, there's a moment that feels like a click, or even a thunderclap, when the narrative (or song, or building, or landscape) takes a new turn. You had no idea that it was coming, but you know it when it's there. With a lot of help from Chris Taylor, author of the terrific *How Star Wars Conquered the Universe*, here's some speculation about what might have happened.

While writing the climactic scene of *The Empire Strikes Back*, Lucas decided, with a burst of inspiration, that Vader should say to Luke, "We will rule the galaxy as father and son." Those words might have jarred Lucas's imagination. They might have produced an "aha," a shiver down the spine. What if Vader meant those words *literally*? As Taylor writes, that could suddenly explain "at a stroke why everyone from Uncle Owen to Obi-Wan to Yoda has been so concerned about Luke's development, and whether he would grow up to be like his father." All this suddenly made a new kind of sense. If the explanation was after the fact—if those concerns originally had nothing at all to do with Darth Vader—well, that's fine. The present often casts a new light on the past, making it somehow different from what we thought. This wouldn't be the first time.

That particular account is just a guess; maybe Lucas came up with the idea that Vader was Luke's father at some earlier stage or in some other way. What really matters is that in the Star Wars series, as in many works of literature, "I am your father" moments and their accompanying shivers are defining. They involve pivotal transitions and reversals of course, which nonetheless maintain (enough) continuity with the previous story, which now changes and gets more interesting. Vader's fatherhood also created a significant challenge for Lucas, because it meant that viewers had to reassess past scenes, sometimes in fundamental ways. If the reassessment produced utter incredulity in the audience—not an "OMG" but a "WTF?"—the "I am your father" moment would not work. In fact it would have backfired, ruining the whole series.

Suppose, for example, that Vader said, "I am your son," or "I am your cat," or "I am Abraham Lincoln," or even, "I am R2-D2." If so, the whole thing would have been a mess. People would have thought, WTF. What was needed was a kind of audience gasp that reflected genuine astonishment, even a moment of stunned disbelief, and then a kind of awed "it all makes sense now." That gasp would show a sense of recognition, a feeling that there is a pattern there after all, even if it was entirely unanticipated.

The best kind of "I am your father" moments produce a sense that everything was foreordained, and everything finally fits. Good mysteries work just that way. Gillian Flynn's *Gone Girl* is one example, and Harlan Coben, the amazing mystery writer, makes an art of these things. A. S. Byatt's magnificent *Possession* has a number of such moments, and on this count, Shakespeare was of course the master of Jedi masters.

If viewers can reassess past scenes in a way that makes the "I am your father" moment seem intelligible and in retrospect even inevitable, the indispensable sense of a coherent narrative is preserved. Of course, narratives, including that of Star Wars, can go in many different directions without losing that sense of coherence. With the best such moments, people could not, in advance, easily imagine the moment—and cannot, in its aftermath, imagine that things could possibly have turned out otherwise.

"FROM A CERTAIN POINT OF VIEW"

Even as it was, the "I am your father" moment in *The Empire Strikes Back* presented Lucas with an acute dilemma: In *A New Hope*, Obi-Wan Kenobi had told Luke that Darth Vader "killed your father." Was he lying? If he was, then Obi-Wan had some serious explaining to do. Why would the sainted Obi-Wan lie to young Luke?

Lucas loves visuals, not plotting, but he found an ingenious solution to this problem. In *Return of the Jedi*, he had Obi-Wan explain: "Your father . . . was seduced by the Dark Side of the Force. He ceased to be Anakin Skywalker and 'became' Darth Vader. When that happened, the good man who was your father was destroyed. So what I told you was true . . . from a certain point of view."

In some circles, this explanation is infamous, a big fat cheat. "From a certain point of view" can be taken as a confession of untruthfulness. Doesn't it sound a bit Sith? If your nation's leader says that, or your spouse, wouldn't you be suspicious? But it's also clever. It makes enough sense to preserve the coherence of the narrative. Of course Obi-Wan's original comments were about a literal killing, not a metaphorical one. But the metaphorical killing provides sufficient coherence; in a way, it's terrific. And if Obi-Wan did not exactly tell the whole truth, well, Luke was young, and maybe he couldn't have handled the revelation.

Chancellor Palpatine, by the way, put Obi-Wan's words into a dark mirror, insisting to young Anakin, "Good is a point of view." The Sith are moral relativists.

LOVING TWINS

What about the fact that Luke and Leia turned out to be twins? In a way, that particular "I am your father" moment turned out to be even more challenging for Lucas's effort to preserve narrative coherence. Mark Hamill himself said, smartly: "This just seemed a really lame attempt to top the Vader thing." But that's too harsh. It didn't top the Vader thing, but it worked, and it solved a whole bunch of problems.

When he wrote *A New Hope* and *The Empire Strikes Back*, Lucas certainly did not think that Luke and Leia were twins. On the contrary—and as the unmistakable sexual tension between the two suggests—he thought that they were *not* siblings. In an interview around 1976, after the release of *A New Hope*, Lucas said, "And who [Leia] ends up with is anybody's guess. I will say that Luke is more devoted to her I think than Han Solo is."

Tellingly, both the early scripts and the casting director noted that Luke was older than Leia (so they couldn't be twins). In the excellent and occasionally steamy novelization of *A New Hope*, written by Alan Dean Foster, Luke responds to Leia's hologram by "savoring the way sensuous lips formed and reformed the sentence fragment." When he first sees her in person: "She was even more beautiful than her image, Luke decided, staring dazedly at her." Here's how the novel ends:

> As he stood awash in the cheers and shouts, Luke found that his mind was neither on his possible future with the Alliance nor on the chance of traveling adventurously with

Han Solo and Chewbacca. Instead, unlikely as Solo had claimed it might be, he found his full attention occupied by the radiant Leia Organa.

Whoa. And after Lucas wrote *A New Hope*, he planned to do a follow-up book or two, saying at the time, "I want to have Luke kiss the princess in the second book. The second book will be *Gone with the Wind* in outer space. She likes Luke, but Han is Clark Gable. Well, she may appear to get Luke, because I want Han to leave."

And of course there is that scene in *The Empire Strikes Back* in which Luke and Leia kiss—and the kiss doesn't look anything like what siblings typically do. Creepy, yes?

True, there were indications that Luke had a sibling—but it was definitely not Leia. When Lucas began *The Empire Strikes Back*, he wrote that Luke had a "twin sister on the other side of the universe— placed there for safety; she too is being trained as a Jedi." So much for the idea that Luke and Leia were to be twins from the beginning.

What Hamill called a "lame attempt" was made in large part because in *The Empire Strikes Back*, Lucas had Yoda say, in response to Obi-Wan's suggestion that Luke was their "last hope," that "there is another." Speaking of that suggestion, Lucas later said that "there are six hours' worth of events before Star Wars, and in those six hours, the 'other' becomes quite apparent, and after the third film, the 'other' becomes apparent quite a bit." It's not clear what exactly Lucas was thinking. He might have been thinking of a sister—but not of Leia, who hardly becomes "apparent" in the six hours that preceded *A New Hope*.

Lucas said that he wrote Yoda's mysterious and intriguing suggestion in part to "enhance the audience's perception of Luke's jeopardy; the story doesn't need him?" Sure, but there was a practical point as well. Mark Hamill might have decided not to return for sequels, which meant that "another" would have to fill his shoes. The suggestion also opened up the possibility of Episodes VII, VIII, and IX, focusing on the twin sister. (By the way, Billy Dee Williams's Lando Calrissian was introduced in part because of a fear that Harrison Ford would refuse to return; Calrissian could fill Han's shoes as the roguish space pirate. Williams was mightily disappointed when Ford came back.)

In fact Lucas did plan, at several stages, to make that third trilogy (though he has not been consistent on this point). But when it came time to write *Return of the Jedi*, Lucas and his actors had little interest in making more movies. They all had had enough. How then to solve the "there-is-another" mystery—and to resolve the romantic triangle that involved Han Solo, Luke, and Leia?

Lucas's solution was to make clear that Leia was the other and that two members of the love triangle were twins. True, the sexual charge between the siblings in *A New Hope* and *The Empire Strikes Back* created quite a problem. As noted, Lucas's solution was to ignore it. And to hammer home (a bit heavy-handedly) the plausibility of their sibling status, he has Leia say, in response to the revelation, "I know; I've always known." Yeah, right. Apparently Lucas hoped that if Leia herself always knew, then the audience might think that their blood relationship was credible, even planned, and not a bizarre, coherence-destroying shift from previous films.

"YOU'VE GROWN WELL, LUKE"

From the originating kernels of Lucas's thinking, the plot could have gone in countless different directions. Most of them would have been worse; the plot would have been less interesting if Obi-Wan had turned out to be Luke's father, or if his father had, in fact, been killed when he was a young child.

The brilliant screenwriter Leigh Brackett produced a late and mostly terrific version of what became *The Empire Strikes Back.* (It is available online.) Sadly, Brackett died shortly after producing that text. She is credited as a coauthor of the actual script, and rightly so. It borrows heavily from her work. It's partly hers, no doubt about that. But what she chose is intriguingly different from Lucas's ultimate text.

In her version, Luke actually meets his late father, a Force Ghost, through the good offices of Obi-Wan Kenobi—and that father is the furthest thing from Darth Vader. As Brackett tells the tale, he is a "tall, fine-looking man," and when Luke sees him, he finds the experience shattering.

Here's the dialogue:

SKYWALKER: You've grown well, Luke. I'm proud of you. Did your uncle ever speak to you about your sister?
LUKE: My sister? I have a sister? But why didn't Uncle Owen? . . .
SKYWALKER: It was my request. When I saw the Empire closing in, I sent you both away for your own safety, far apart from each other.
LUKE: Where is she? What's her name?
SKYWALKER: If I were to tell you, Darth Vader could get that information from your mind and use her as a hostage. Not yet, Luke. When it's time . . . Luke, will you take from me the oath of a Jedi Knight?

Then Luke repeats: "I, Luke Skywalker, do swear on my honor, and on the faith of the brotherhood of knights, to use the Force only for good, turning always from the Dark Side, to dedicate my life to the cause of freedom and justice. If I should fail of this vow, my life shall be forfeit, here and hereafter."

The vow is silly, as is the whole scene, and Lucas's I-am-your-father twist made everything less gee-whiz, and edgier, and just better. (Lucas visited the Dark Side.) It's worth asking, though, whether Lucas always made the right choice. Some of the imaginable different directions would have been good ones. We shouldn't join the chorus, bashing the ambitious, visually spectacular, and underrated prequels, but this is certainly true for the less-than-perfect love scenes between Anakin and Padmé:

> **ANAKIN SKYWALKER:** You are so . . . beautiful.
> **PADMÉ:** It's only because I'm so in love.
> **ANAKIN SKYWALKER:** No, it's because I'm so in love with you.
> **PADMÉ:** So love has blinded you?
> **ANAKIN SKYWALKER:** [laughs] Well, that's not exactly what I meant.
> **PADMÉ:** But it's probably true.

Also not ideal, from Padmé: "Hold me, like you did by the lake on Naboo; so long ago when there was nothing but our love. No politics, no plotting, no war."

But we might be able to identify improvements on some of the much-admired scenes. On those, it's not easy to outdo Lucas—in the first trilogy, he felt the Force, big-time—but it would be amazing if he always found the very best path. In Brackett's telling, Luke and Leia are not brother and sister, and Leia has love scenes with both Luke and Han before settling on the latter, leaving Luke as a mature and resigned solo. That could have been interesting.

Perhaps Luke might have had a twin sister on another planet, who could have electrified the plot. If something like that had happened and become canonical, we might have said, "Yuck!" or "How stupid and ridiculous" if we had been told about the Luke-and-Leia-are-siblings twist that Lucas actually chose.

Wise words from the musician Skrillex: "The future is an accident. It's an accident because you explore. . . . You can't see it—you just have to go somewhere you haven't been before." And of course J. J. Abrams made a series of distinctive choices for *The Force Awakens;* he could have gone in many other directions. (He almost did.) After the writing of that film, the plots of Episodes VIII and IX had to be chosen as well. They were hardly foreordained. About Abrams, a nice point from film critic Anthony Lane: "I hate to say it, but he's a critic—as all creators, and especially re-creators, must necessarily be."

"I DON'T LIKE THAT AND I DON'T BELIEVE THAT"

Here is a classic example of choice-making in the construction of the Star Wars narrative, in real time. It's from a crucial moment in the actual writing of *Return of the Jedi*. It involves a sharp artistic disagreement between Lucas, at the very height of his extraordinary powers, and Kasdan, one of the most brilliant (and I think deep) screenwriters of the last half century.

It's a clash of two Jedi masters, with radically different visions for the film:

KASDAN: I think you should kill Luke and have Leia take over.

LUCAS: You don't want to kill Luke.

KASDAN: Okay, then kill Yoda.

LUCAS: I don't want to kill Yoda. You don't have to kill people. You're a product of the 1980s. You don't go around killing people. It's not nice.

KASDAN: No, I'm not. I'm trying to give the story some kind of an edge to it. . . .

LUCAS: By killing somebody, I think you alienate the audience.

KASDAN: I'm saying that the movie has more emotional weight if someone you love is lost along the way; the journey has more impact.

LUCAS: I don't like that and I don't believe that.

KASDAN: Well, that's all right.

LUCAS: I have always hated that in movies, when you go along and one of the main characters gets killed. This is a fairy tale. You want everybody to live happily ever after and nothing bad happens to anybody. . . .
The whole point of the film, the whole emotion that I am trying to get at the end of this film, is for you to be real uplifted, emotionally and spiritually, and feel absolutely good about life. That is the greatest thing that we could possibly ever do.

In my view, Lucas wins the argument by a knockout. Precious words: "I don't like that and I don't believe that." The words are precious partly because of how they are ordered. "Not liking" precedes, and helps account for, "not believing." If you don't like something, you're going to be inclined to disbelieve it. This is the psychologists' idea of "motivated reasoning." (Also good: "You don't go around killing people. It's not nice.")

I don't like what Kasdan says, and I don't believe it, either.

But Yoda does end up dying. (Kind of.) And of course Kasdan finally got his wish in *The Force Awakens*, with the murder of Han Solo. As he put it in 2015: "I was always lobbying to kill somebody major because that gave gravitas to the story. If everyone always comes out fine then there was no danger. It should cost us." Chose *that*, Kasdan did.

With reverence and respect: Wrong choice! After seeing Han's death, one of my friends, plainly distraught, announced, "I am done. I am not seeing any more of these movies." Right after the movie, she spent ten minutes in the bathroom, crying. I'm not done, and I'm seeing every single one, but I have always hated that in movies, when you go along and one of the main characters gets killed.

PARSECS

And then there is, of course, Han Solo's immortal line about the Millennium Falcon: "It's the ship that did the Kessel Run in less than twelve parsecs." With its irresistible specificity, the phrase seems recognizable, even familiar: "less than twelve parsecs" sounds real, even though a "parsec" is a unit of distance rather than time, so without some fancy footwork, the line doesn't make a lot of sense. At the same time, Solo's language is impossibly foreign. What's "the Kessel Run"?

Delivered with smug self-satisfaction by Harrison Ford, the line captures much of what makes the series work. It is the incarnation of what Lucas called the most characteristic feature of his films: "effervescent giddiness." (He added that he found that puzzling, because he's not at all like that as a person.) The novel version isn't nearly as good: "It's the ship that made the Kessel run in less than twelve standard timeparts!"

Star Wars may be a grain of sand, but it does contain a whole world.

THE MOVIE NO ONE LIKED

An Expected Flop Becomes
the Defining Work of Our Time

There's this giant guy in a dog suit walking around. It was ridiculous.

—HARRISON FORD

Was the Star Wars series bound to succeed? Was it all a matter of destiny? Surely so?

Right out of the gate, *Star Wars* (now known as *A New Hope*) was a spectacular hit. On its opening day, May 27, 1977, it was shown in just thirty-two theaters, but it broke records for nine of them, including four of the five New York theaters in which it was screened. Even though the movie premiered on a Wednesday, its single-day total was $254,809, or $8,000 per location. At Mann's Chinese Theatre, in Hollywood, the single-day total alone was $19,358, while Manhattan's Astor Plaza brought in $20,322.

True, it did not quite win the box-office sweepstakes on its opening weekend; *Smokey and the Bandit* earned $2.7 million, to $2.5 million for *A New Hope*. But the immortal *Smokey* was shown on a whopping 386 screens, and with a grand weekend total of 43, *A New Hope* didn't have much of a chance.

The movie continued to be a sensation throughout the summer. Entire towns mobilized, seemingly en masse, in order to see it. Just one example: within a few months of its release, fully half of the population of Benton County, Oregon, had seen the film. (What happened to the other half?) As the amazement and exhilaration spread, its popularity grew in its initial months, finally peaking in mid-August, when it was showing at approximately 1,100 theaters across the country. Its appeal sustained itself over time. Some 42 theaters screened the movie continuously for *over a year*. Across the United States, theaters were forced to order new prints because the old ones were literally being worn out.

Of course, *A New Hope* was a smashing financial success. By September, it had become Twentieth Century Fox's most successful film, ever. As a direct result of the movie's performance, the studio's stock skyrocketed: it jumped from $6 a share to nearly $27 a share in the immediate aftermath of the movie's release. In just a few months, *A New Hope* surpassed *Jaws* to become the highest-grossing film of all time. When its first theatrical run finally ended, it had made $307 million.

That's 240 percent of the earnings of the second-highest grossing film of 1977, *Close Encounters of the Third Kind*, which earned $128 million. It's also roughly six times the earnings of the year's fifth-highest earner, *A Bridge Too Far*—which brought in $50.8 million—and approximately eighteen times the earnings of *Kingdom of the Spiders*, which came in tenth at the box office with $17 million. If we include its rereleases and adjust for ticket price inflation, *A New Hope* has made an estimated $1.55 billion at the box office. By way of comparison, this number exceeds *Avatar*'s adjusted earnings by over $600 million; in terms of GDP, it trumps Samoa by about $700 million. (In inflation-adjusted earnings, only *Gone with the Wind* tops *A New Hope*, and they're not that far apart. *A New Hope* is comfortably ahead of *The Sound of Music*, *E.T.*, *Titanic*, *The Ten Commandments*, and *Jaws*.)

The five Lucas sequels and prequels also enjoyed terrific levels of success. *The Empire Strikes Back* went on to make $209 million during its first box-office run, and all of the later Lucas films made well over $200 million in their initial theatrical releases. *The Phantom Menace* is probably the worst of the lot, but of the two Lucas trilogies, it had the highest unadjusted gross. Once you adjust for inflation (as you should), it still ranks an impressive eighteenth of all time, just two spots below *Return of the Jedi* and five below *The Empire Strikes Back*. (And *The Phantom Menace* is better than you think; a lot of the scenes crackle with imagination. Remember the pod race? The awesome fight with Darth Maul?)

If *The Force Awakens* is any gauge, Star Wars is going to be a financial Jedi for a long time. As of early 2016, it ranked as the eleventh most successful movie of all time in terms of adjusted income, and among Star Wars movies, only *A New Hope* did better. In its opening weekend, *The Force Awakens* brought in roughly $517 million worldwide. The North American numbers topped $238 million, with $120 million from its first full day and $57 million from its opening night alone. To put those numbers in context, the previous North American record was *Jurassic World*'s $208 million. *The Hobbit: An Unexpected Journey* previously held the record for biggest December opening weekend with $87.5 million, and *Avatar*—the highest-grossing film in history—took in a comparatively ordinary $85 million domestically during its opening weekend. *The Force Awakens* earned $1 billion more quickly than any movie in history—just twelve days. In terms of unadjusted income, it became the highest-grossing domestic film of all time a mere eight days later.

All these are just numbers, of course. In terms of the culture, the figures don't come close to capturing the impact of the series. *Avatar* was a massive economic success, and it was really good, but can you recall even one line or scene? You might remember a few things from *Titanic* or *The Wizard of Oz* or *Gone with the Wind*—but frankly, my dear, I don't give a damn. We're not in Kansas anymore; it is Star Wars that has become king of the world.

Around the world, presidents know about it, and so do senators and Supreme Court justices, and so does your kid, and so do your parents. If you want to bond with someone you don't know, try talking about Star Wars. It's a lot better than the weather, and it's likely to work. (A revealing exception: I was recently at a dinner with Syrian refugees, a wonderful family with five children. Not one of them had heard of Star Wars.) In the aftermath of the release of *The Force Awakens*, I was asked to come to Copenhagen to give several talks about public policy and regulation; my hosts also wanted me to speak about Star Wars. At holiday parties in New York in 2015, the main topic was not the presidential race or Hillary Clinton or the Affordable Care Act or Russia. The Force had awakened.

"NOBODY LIKED IT"

But there's an irony here, and a major puzzle, too. In the early stages, Lucas reports, "*nobody* thought that it was going to be a big hit." When *A New Hope* was released, a lot of insiders thought that they had a dud on their hands. Throughout its production, there was "basic apathy toward the project with Fox," and many executives had "little faith in the film or its director." Wild but true: They "hoped a lot of times that it would just go away."

It's revealing that when Lucas and his team started to run out of cash, Lucas had to pay directly out of the money he had made from *American Graffiti* (which was also a wholly unanticipated hit). Without that infusion of personal cash, the whole project might well have collapsed. Nor was the near-universal negativity solely a product of trepidation about the unusual nature of the whole project. (Droids? The Force? Some old guy named Obi-Wan? Played by Alec Guinness? Lightsabers?) After the board at Fox finally saw a rough cut, there was "no applause, not even a smile. We were really depressed."

Even during the last stages, Lucas himself "didn't think the film was going to be successful." Most people at the studio agreed; "the board didn't have any faith in it." As evidence of its lack of faith, the studio saw fit to show just one winter trailer for the film, at Christmas; it was shown only once more, during Easter.

Astonishingly, Fox seemed to think that the movie wasn't even worth the celluloid on which it was printed—literally. The studio made fewer than one hundred prints, which caused a terrible problem once the crowds demanded to see it. Far more optimistic than most, Lucas himself projected that because young people might like it, it could earn $16 million, about as much as the average Disney movie. He claimed that the chances of its doing much better than that were "a zillion to one."

"GIANT GUY IN A DOG SUIT"

Even after its smashing success, Lucas said, "I expected to break even on it, I still can't understand it." His then wife and close collaborator,

Marcia Lucas, thought that Martin Scorsese's *New York, New York* (which she also helped edit) would do better.

For their part, movie theaters, whose business is to know what people will like, responded very cautiously. Fox hoped for advance guarantees of $10 million but got only a fraction of that: a humiliating $1.5 million. Its most promising film of the summer, it thought, was *The Other Side of Midnight*, and to coerce interest in Lucas's film, the studio warned theaters that if they didn't take *A New Hope*, they wouldn't get that film, either.

The whole marketing enterprise might have collapsed if not for the tireless efforts of Charley Lippincott, a friend of Lucas, who had a lot of faith in the movie. Lippincott promoted it aggressively and helped it to get it into those admittedly paltry thirty-two theaters—one of which was the prestigious and large Coronet in San Francisco. As it turned out, his success with the Coronet really mattered.

Immediately after *A New Hope* opened, Lucas and his wife went on vacation in Hawaii. They wanted to be away not only because they needed a vacation, but also because as the reviews came in, they feared that Lucas "had just released a flop." In remote Hawaii they could escape what "Lucas was certain was going to be a disaster." Years later, he reported that even his friends "didn't have any faith in it either. And the [studio] board didn't have any faith in it. . . . Nobody liked it."

The actors agreed. Anthony Daniels (C-3PO, of course) said, "There was a general atmosphere on the set that we were making a complete turkey." Harrison Ford noted, "There's this giant guy in a dog suit walking around. It was ridiculous." David Prowse, who played Darth Vader (though of course James Earl Jones did the all-important voice), noted, "Most of us thought we were filming a load of rubbish." Mark Hamill observed, "I remember thinking, it's really hard to keep a straight face doing this stuff. Alec Guinness sitting next to a Wookie—what's wrong with this picture?" Years afterward, Carrie Fisher recalled, "The film wasn't supposed to do what it did—nothing was supposed to do that."

The sound designer, Ben Burtt, thought the movie might be a success for a few weeks: "The best I could imagine was that we would get to have a table at next year's Star Trek convention." Even after the massive opening crowds, Lucas said, "Science fiction films get this little group of sci-fi fans. They'll come to anything in the first week. Just wait." As one film scholar summarized the evidence, "no one predicted" the critical admiration and viewer fanaticism that followed release of *A New Hope*.

Why didn't anyone see what was coming? Aren't movie studios, and the experts who staff them, supposed to be good at that sort of thing?

SECRETS OF SUCCESS

Was Star Wars Awesome, Well-Timed, or Just Very Lucky?

Ultimately, we're all social beings, and without one another to rely on, life would be not only intolerable but meaningless. Yet our mutual dependence has unexpected consequences, one of which is that if people do not make decisions independently—if even in part they like things because other people like them—then predicting hits is not only difficult but actually impossible, no matter how much you know about individual tastes.

—DUNCAN WATTS

Why do some products (movies, books, TV shows, songs, politicians, ideas) succeed, and why do others fail? In answering that question, we're going to focus on the Star Wars phenomenon, and at the same time try to draw general lessons about success and failure—not only for products of all kinds but for people of all sorts as well.

For your consideration, here are three hypotheses.

QUALITY

The first is that *intrinsic quality is what determines success*. Because of its energy, spirit, and originality, Star Wars could not possibly have failed. It's too awesome. In advance, hardly anyone saw that. Steven Spielberg was an exception; he loved it from the start. Spielberg is right about almost everything, and he was right about *A New Hope*.

Lawrence Kasdan captures its awesomeness this way: "it's fun, it's delightful, it moves like a son of a bitch, and you don't question too much." Great things inevitably rise to the top (especially if they move like an SOB). With movies, books, music, and art, there's no mystery about what succeeds and what fails. Shakespeare, Dickens, Michelangelo, Mozart, Frank Sinatra, the Beatles, Taylor Swift—all these were destined for success. You can't imagine a world in which *Hamlet* or *King Lear* tanked. (Perhaps you can imagine one in which people don't like Taylor Swift, but if so, I feel sorry for you.) Quality is necessary, and it is also sufficient.

SOCIAL INFLUENCES

The second hypothesis is that *while intrinsic quality is necessary, it really isn't enough; a successful movie, book, or work of art requires social influences and echo chambers to get people excited*. There's a lot of tremendous stuff out there, and no one has ever heard of most of it. Some sparks turn into big fires, and others flame out. What often matters is whether a bandwagon effect is generated, so that people start to like it *because they think other people like it*. George Lucas produced a tremendous movie, or two, or maybe even four, but he was also supremely lucky, in the sense that a massive bandwagon effect got going on his behalf. (J. J. Abrams had a much easier path with *The Force Awakens*, because the brand was already a brand.) And with a little twist of fate, maybe no one would have heard of Lucas, or Dickens or Sinatra, or even Mozart or Shakespeare or Swift.

To test this claim, of course, we would need to be pretty specific about terms like "social influences" and "echo chambers." But the basic idea is that even if you have something great, your product might not go anywhere. (Look around, and you'll see.)

TIMING

The third hypothesis is that what matters is *the relationship between the product and the culture at the particular time that it is released*. Some artists and movies hit a particular cultural nerve, and that's both necessary and sufficient for success. Sure, they're terrific, but without good timing, they'd fail. Other artists and movies are fabulous, but the culture isn't ready for them, or their time has passed, so they tank. What you need is *cultural resonance*.

On this view, Lucas definitely hit the right nerve. The first release of Star Wars wasn't originally called *A New Hope*, but everyone understood it exactly that way, because that's exactly what it was. Lucas intended it for young people—for fourteen-year-olds or under—and the movie connected with the kid in each of us, who needed the attention, and some optimism and exhilaration, at just that time. After the terrible tumult of the 1960s—the assassinations of two Kennedys, of Martin Luther King Jr., of Malcolm X—a new hope was exactly what people wanted. Lucas handed it to them on a silver spaceship.

For some, it helped that the Empire could be seen as the United States, or at least the Nixon administration (as Lucas himself suggested).

But for others, it didn't hurt that the movie was released in the closing stages of the Cold War, where the Empire could easily be associated with the Soviet Union. (Was it a coincidence that in 1983 Ronald Reagan called it "an evil empire"?)

Similarly, it's no accident that Harry Potter and *The Hunger Games* took off in the first and second decades of the twenty-first century. After the attacks of September 11, 2001, people wanted entertainment that played on widespread anxieties about evil (Voldemort as Osama bin Laden?) or that triggered dreams about heroic freedom fighters. *Star Wars*, the Harry Potter films, and *The Hunger Games* had something in common: they fit with the zeitgeist. Harry Potter showed that with a little magic, the good guys could triumph. *The Hunger Games* managed to combine science fiction and adventure (conventional boy stuff) with a strong sense of romance (conventional girl stuff), and it also played on anxieties about technology and surveillance. Successful artistic works may be great, or may be okay, or may be rotten, but they succeed only if they resonate.

Which hypothesis is right?

SUGAR MAN

To answer that question, let's turn to another film, one that has nothing to do with spaceships or droids. In 2012, the Oscar for best documentary was awarded to *Searching for Sugar Man*. The film focused on an unsuccessful Detroit singer-songwriter named Sixto Rodriguez, also known as Sugar Man, who released two albums in the early 1970s. He's good, even terrific, but you probably haven't heard of him. Almost no one bought his albums, and his label dropped him.

Reasonably enough, Rodriguez stopped making records and sought work as a demolition man. His two albums were forgotten. Rodriguez joined the ranks of countless people who have tried to fulfill some kind of artistic dream, only to discover that the competition is fierce, and very few can survive. A family man with three daughters, Rodriguez was hardly miserable. But working in demolition, he struggled.

Having abandoned his musical career, Rodriguez had no idea that he had become a spectacular success in South Africa—a giant, a legend, comparable to the Beatles and the Rolling Stones. People said his name slowly and with awe, even reverence: "Rodriguez." Describing him as "the soundtrack to our lives," South Africans bought hundreds of thousands of copies of his albums, starting in the 1970s. His South African fans speculated about his mysterious departure from the musical scene. Why did he suddenly stop making records? According to one rumor, he burned himself to death onstage. *Searching for Sugar Man* is about the contrast between the failed career of Detroit's obscure demolition man and the renown of South Africa's mysterious rock icon.

The film is easily taken as a real-world fairy tale, barely believable, a story so extraordinary that it gives new meaning to the phrase "you couldn't make it up." But it is a bit less extraordinary than it seems, and it offers significant lessons about cultural success and failure—and about something that really gave a boost to *A New Hope*.

Let's begin by conceding that some kind of quality is usually necessary; if Rodriguez had written really awful songs, he wouldn't have made it in South Africa. But often quality is far from enough. (So much for the first hypothesis.) For most things that have a large cultural impact, social dynamics are the crucial factor, and you need a bit of luck as well as skill to move those dynamics in your favor. (The Rebellion did just that, in *Return of the Jedi*, and so did the Emperor, in *Revenge of the Sith*.) Who is conveying enthusiasm to whom, and how loudly, and where, and exactly when? The answers can separate the rock icon from the demolition man, and mark the line between stunning success and crashing failure. An understanding of those dynamics tells us a lot about why success and failure can be impossible to predict.

RESONATING WITH THE CULTURE

At this stage, you might want to turn to the third hypothesis and venture a point about South African culture in the 1970s. Maybe Rodriguez, with his songs of protest and freedom and exclusion, struck a particular chord in a nation that was riven by the debate over apartheid. Maybe Rodriguez's audience there—overwhelmingly young and white—was distinctly, even uniquely ready for him.

Probably not. There were lots of good singers in the late 1960s and early 1970s who spoke of protest and freedom and exclusion. (In that period, almost every singer did.) In South Africa, Rodriguez was the only one who broke through. How come? It's not easy to say, except perhaps in retrospect, that he and South Africa were especially made for each other.

This point holds for many products that turn out to do spectacularly well. After the fact, we construct just-so stories, and they're entirely plausible. After the 1960s, the world was ready for *Star Wars*, and after the attacks of 9/11, the world was ready for *Harry Potter*, and also for *Mad Men* and *The Hunger Games* and Taylor Swift. After the financial crisis, *Gone Girl* was bound to become a bestseller, and *Mad Max* was bound to be remade and to do great.

True, those last two don't make any sense at all—but that's exactly my point. We can always come up with some account of why everything that happened was bound to happen, but who knows whether it's right?

Was *A New Hope*, then, a lot like Rodriguez in South Africa—a beneficiary of favorable social dynamics, which catapulted it to stupendous success? With a little twist or turn, could *A New Hope* have been like Rodriguez in the United States—a victim of unfavorable social dynamics, which might have made it a flop, one of a large set of failed (but sometimes great) science fiction movies or TV shows? (A fantastic example: *Awake*, a one-season TV show from 2012. It never was a hit, but it's brilliant, and fun, too. Wow. Why hasn't it found its South Africa? See it!)

THE MUSIC LAB

A few years ago, social scientists Matthew Salganik, Duncan Watts, and Peter Dodds became intrigued by the question of cultural success and failure. Their starting point was that like the hapless people who thought that *A New Hope* would flop, those who sell books, movies, TV shows, and songs often have a lot of trouble predicting what will succeed. Nonetheless, some of those products are spectacularly successful, far more so than the average—which suggests, very simply, that those that succeed must really be a lot better than those that don't. If they are so much better, then why are predictions so difficult?

Here's some concrete evidence of the difficulty of foresight, even among experts: In 1996, J. K. Rowling's manuscript for her first Harry Potter book was rejected by no fewer than twelve publishers. Eventually, Bloomsbury agreed to publish it, but with a very small advance (£1,500). To date, the series has sold more than 450 million copies worldwide. Why couldn't even one of those twelve publishers have figured that out? Why wasn't there a huge bidding war?

To explore the sources of cultural success and failure, Salganik and his coauthors created an artificial music market on an actual website, which they called the Music Lab. The site offered people an opportunity to hear forty-eight unknown songs by unknown bands. One song, for example, by a band called Calefaction, is "Trapped in an Orange Peel." (I agree, that's the worst title ever.) Another, by Hydraulic Sandwich, is "Separation Anxiety."

The experimenters randomly sorted half of about 14,000 site visitors into an "independent judgment" group, in which they were invited to listen to brief excerpts, to rate songs, and to decide whether to download them. For 7,000 or so visitors, Salganik and his coauthors could get a clear sense of what people really liked best. The other 7,000 visitors were sorted into a "social influence" group, which was exactly the same except in just one respect: they could see how many times each song had been downloaded by other participants.

Here's the ingenious part of the experiment: People in the social influence group were also randomly assigned to one of eight subgroups, *in which they could see only the number of downloads in their own subgroup.* In those different subgroups, it is inevitable that as a result of random factors, different songs would attract different initial numbers of downloads. For example, "Trapped in an Orange Peel" might attract strong support from the first few listeners in one subgroup, whereas it might be a big bust in another.

The research question was this: would the initial numbers matter to where songs would ultimately end up on the researchers' hit parade? You might expect that quality would always prevail—that the popularity of the songs, as measured by their download rankings, would be roughly the same in the independent group and in all eight of the social influence groups. That expectation fits with the first hypothesis: *A New Hope* was destined to succeed.

But that isn't what happened—not at all. "Trapped in an Orange Peel" could be a huge hit or a miserable flop, depending on whether a lot of other people initially downloaded it, and were seen to have done so. In short, everything turned on initial popularity. Almost any song could end up like Rodriguez in South Africa or Rodriguez in the United States, *depending on whether or not the first visitors liked it*. Importantly, there is one qualification: the songs that did the very best in the independent group rarely did very badly, and the songs that did the very worst in the independent group rarely did spectacularly well. But otherwise, almost anything could happen.

Here's a cautious, modest reading of these findings. Some products really are destined for success, and others really are destined for failure. If a song is truly sensational, it will be a hit. Mozart, Shakespeare, and Dickens were bound for success (and the same might be true for *A New Hope*). If a song sounds horrible, it's going to flop. If you have no talent, forget about it. But within a wide range, songs can do very well or very poorly, and within that range, you just can't predict. For music, everything seems to depend on social influences—on something like a real-life attack of the clones.

This is a plausible reading of the Music Lab experiment, but I think that it is far too cautious (and so do Salganik and his coauthors). Sure, terrible songs, movies, and books are unlikely to succeed. But maybe the best ones are not destined for success. Almost nothing is. After all, the Music Lab experiment itself was tightly controlled. It had had just forty-eight songs. In real markets there are countless more. And in those real-world markets, media attention, critical acclaim, marketing, and product placement play a big role. Recall the indispensable work of Charley Lippincott, who gave *A New Hope* a big initial boost.

We haven't reached a final conclusion, but my second hypothesis—that the success of *A New Hope* depended on luck, in the form of favorable social influences—is starting to look pretty reasonable.

CUCKOO

Let's go back to J. K. Rowling and consider the saga of *The Cuckoo's Calling*, a crackling detective novel with a big heart, published in 2013 by an unknown author, Robert Galbraith. The novel got some excellent reviews, but it didn't sell well. A critical success but a commercial failure, it looked as if it would join the ranks of the many literary Rodriguez types—excellent, maybe even better than that, but unable to hit the big time. Maybe Galbraith would quit writing and become a demolition man.

After a while, however, a little information was released to the public: "Robert Galbraith" was, and is, J. K. Rowling!

In short order, *The Cuckoo's Calling* vaulted to the bestseller list. It deserved it, but it couldn't possibly have gotten there without the magic of Rowling's name. Of course that's not quite the Music Lab. *The Cuckoo's Calling* didn't bust out because early adopters liked it. But it's similar. Whatever its quality, the novel needed some kind of social boost, and that magical name did the trick. (Note also that if the novel had been terrible, it would have struggled, even with the name; quality was necessary.) Galbraith/Rowling followed *The Cuckoo's Calling* with two other novels (as of early 2016). They're both terrific, and everyone should read them. But they also were big hits, as they wouldn't have been without Rowling's name.

Still unconvinced? Consider a mischievous experiment from Salganik and Watts. That experiment drew on their work with the Music Lab—but they inverted the actual download figures, so as to make people think that the least popular songs were the most popular, and the most popular songs the least. If quality is the real driver, you might expect that the worst songs would eventually plummet, and the best ones would eventually rise to the top. Surely something like that would happen?

Not at all. With the inversion, Salganik and Watts could transform the worst songs into big winners. They could also make most of the top songs into big losers. Here, as in their principal experiment, the lesson is that people pay a lot of attention to what other people appear to like, and information about popularity can make all the difference. The wrinkle is that the very best songs (again, as measured by actual popularity in the control tradition) always ended up doing pretty well; social influences could not keep them down.

That's highly suggestive, but it's reasonable to wonder whether that's true in the real world. Suppose that someone made an awful movie named *Star Cars*, and somehow succeeded in getting a lot of early buzz. Even in the short run, that movie is going to crash. Robert Galbraith produced a terrific novel, but until Rowling's authorship came out, it didn't sell so good.

"WRIT IN WATER"

If any literary figures are great, surely the short list includes William Wordsworth, John Keats, Jane Austen, and William Blake. Equally surely, it does not include George Crabbe, Robert Southey, Barry Cornwall, Leigh Hunt, and Mary Brunton. But H. J. Jackson's important study of literary reputation strongly suggests that even for the greatest of the great, accident, contingency, and luck play a massive role.

In terms of perceived quality, Wordsworth, Crabbe, and Southey were grouped together during their lifetimes. The same is true of Keats, Cornwall, and Hunt, and also of Austen and Brunton. If you asked their contemporaries which names on the list would be famous in the twenty-first century, there would be no consensus in favor of Wordsworth, Keats, and Austen.

Jackson notes that Keats may well count as the most loved poet of all time—but at the time of his death, he believed that he had utterly failed in his somewhat desperate quest for literary fame, leaving instructions that his gravestone have no name, but only these pitiful words: "Here lies one whose name was writ in water." In Keats's time, Cornwall was far more successful; he was regarded as the great poet while Keats was met with "indifference or hostility."

Tracing Keats's improbable rise to prominence decades after his death, Jackson writes, "It seems that his reputation was dependent less on the efforts of particular individuals than on groups, overlapping networks of like-minded acquaintances starting up on a small scale, the collective chatter that later becomes the buzz of fame." (Was Keats like Rodriguez in South Africa?) In terms of pure poetic quality, Cornwall's virtues and vices greatly overlap with those of Keats. (Was Cornwall like Rodriguez in the United States?)

Jackson's remarkable conclusion, which she makes quite plausible, is that "[a]s far as reputation is concerned, the differences between them are largely personal and accidental." At the very least, it is necessary to come to terms with the "conundrum of Barry Cornwall's success with the same audience that spurned Keats." Indeed, their contemporaries put Cornwall far above Keats—and the now-obscure Hunt ranked above them both. And if we are interested in professional opinions, we will find that Wordsworth, Samuel Taylor Coleridge, and Lord Byron all ranked Cornwall highest of the three.

In terms of changing reputations over time, Jackson places a particular emphasis on echo chamber effects, which can consolidate a writer's image. In their time, Mary Brunton and Jane Austen were about equally well regarded, but the former of course faded into obscurity. Jackson urges, and demonstrates, that what "happened to Brunton—the gradual fading and extinction of her name—could easily have happened to Austen." The long-obscure (and fabulous) Blake himself was a beneficiary of a highly improbable and complex recovery project, barely rescuing him from literary oblivion. In his time, his works "were almost unknown to his contemporaries." Jackson's conclusion is that "long-term survival has depended more on external circumstances and accidental advantages than on inherent literary worth."

Maybe the most famous literary figures are, in fact, greater than those who are unknown. But maybe not. With a little push or shove, the literary canon could feature Crabbe, Hunt, and Brunton. Maybe they really are Rodriguez in Detroit, or the unknown author of *The Cuckoo's Calling*.

WHAT CROWDS DO

Let's go back to the Music Lab and try to explain what happened there. As we'll see, the explanation bears directly on the success of the Star Wars series.

Network Effects

Some things can be enjoyed all by yourself. You might like a walk in the sun, or a cup of coffee, or a quick swim, even or perhaps especially if you are alone. Other pleasures are guilty. You might love a silly television show, and you really don't want to watch it with anyone else. But sometimes *the value of a good depends on how many other people are using it.* It isn't a lot of fun to have a telephone if you are the only person in the world who has a telephone. People use Facebook because a lot of people use Facebook. If Facebook hadn't been able to build a network, it would have failed. *Network effects* exist when value increases with the number of users.

Star Wars isn't exactly a telephone, but it benefits greatly from network effects: it is one of a number of cultural goods of which people think they should be aware. Quite apart from its intrinsic merits, it's good to know about, so that you can talk to others about it. It might not be a lot of fun to stare blankly when someone makes a knowing reference to Kylo or Han or Emperor Palpatine or Rey, or even to Chewbacca. If you think that people like Star Wars and focus on it, you might join them for one reason above all: You don't want to be left out. You want to be part of the group.

Arion Berger notes that "it's fun to participate in some cultural swoon," and that's exactly what Star Wars has been. Here's how Ann Friedman puts in: "Ultimately, I realized, I was going to see *The Force Awakens* because all my friends were going to see it, and everyone else's friends were going to see it, too. I was in the grip of an increasingly rare phenomenon: A true mass-cultural event."

Within twenty-four hours of release of the first teaser trailer for *The Force Awakens*, it had been viewed more than 88 million times.

That's an all-time record. You can be sure that a lot of viewers were interested less in seeing the trailer itself than in being able to discuss it with others. As Berger notes, Star Wars is "simultaneously a cult artifact and a staggeringly popular phenomenon." In an era of cultural fragmentation, that's a neat trick, and socially even precious. In such an era, people like it; they might even need it.

Informational Cascades

An important explanation of cultural success, one that bears directly on both *A New Hope* and *The Force Awakens*, goes by an unlovely name: *informational cascades*. There are just too many products out there, and too many ideas as well. None of us is able to sort through them all. Why, exactly, did you pick up that novel you recently finished? Why are you upbeat or even excited about that particular politician? Typically we rely on what other people think, or seem to think. And when a lot of people think or do something, we will probably be affected.

To see how informational cascades work, imagine that seven people are in a reading group, trying to decide which book to try next. Assume that group members are announcing their views in sequence. Each person attends, reasonably enough, to the judgments of others. Leia is the first to speak. She suggests that the new Robert Galbraith book is the one to try. Finn now knows Leia's judgment; he should certainly go along with Leia's account if he is also excited about that book. But suppose that he doesn't really know. If he trusts Leia, he might simply agree: let's try Galbraith.

Now turn to a third person, Luke. Suppose that both Leia and Finn have said that they want to try Galbraith, but that Luke's own view, based on limited information, is that the book is not likely to be very good. (Luke is wrong, because Galbraith/Rowling is fabulous, but let's bracket that point.) Even if Luke has that view, he might well ignore what he knows and just follow Leia and Finn. The reason isn't that Luke is a coward. It is likely, after all, that both Leia and Finn have reasons for their enthusiasm. Unless Luke thinks that his own information is really better than theirs, he should follow their lead.

If he does, Luke is in a cascade. True, Luke will resist if he has sufficient grounds to think that Leia and Finn are being foolish. But if he lacks those grounds, he is likely to go along with them.

Suppose that Han, Chewbacca, Biggs, and Rey are now expected to express their views. If Leia, Finn, and Luke have all said that Galbraith is the author to read, each of them will probably reach the same conclusion even if they have some reason to think another choice would be better. The trick in this example is that Leia's initial judgment has started a process by which a number of people are led to participate in a cascade, leading the whole group to opt for Galbraith. (As we will see, political movements, including rebellions and resistance, often start in this way.)

This is, of course, a highly artificial example. But the basic point should be plain. People learn from others, and if some people seem to like something, or want to do something, you might like or do the same. At least this is so if you do not have reason to distrust them and if you lack a good reason to think that they're wrong.

Informational cascades tend to be fragile, and if they are leading people in bad directions, they are usually broken. If our little reading group converges on a bad book, they'll find out soon enough, and if they're talking to other people, other groups won't be likely to read that book. Word of mouth spreads rapidly, which makes cascades both inevitable and highly vulnerable.

A New Hope very much benefited from an informational cascade, but if it hadn't been terrific, it would have been a mere fad.

Reputational Cascades

Sometimes people pay attention to the views of others because they want to know what's good. But sometimes what they really want is for other people to like them, or at least not to dislike them. That's why they follow the views and actions of others. If most people are enthusiastic about a new song or movie, they might show enthusiasm, too—or at least listen or look. The underlying point here involves conformity.

In a reputational cascade, people think that they know what is right, or what is likely to be right, but they nonetheless go along with the crowd in order to maintain the good opinion of others. Suppose that Boba suggests that the new *Star Cars* movie is spectacular, and that Kylo concurs with Boba, not because he actually thinks that Boba is right, but because he does not wish to seem, to Boba, to be some kind of fool or idiot. If Boba and Kylo say that the new *Star Cars* movie is great, Rey might not contradict them publicly and might even appear to share their judgment—not because she believes that judgment to be correct, but because she does not want to face their hostility or lose their good opinion.

It should be easy to see how this process might generate a cascade on behalf of *Star Cars*. Once Boba, Kylo, and Rey offer a united front on the issue, their friend Poe might be reluctant to contradict them even if he thinks that they are wrong. The apparently shared view of Boba, Kylo, and Rey carries information; that view might be right. But even if Poe has reason to believe that they are wrong, he might not want to take them on publicly. His own silence will help build the reputational pressure on those who follow.

FAMOUS FOR BEING FAMOUS

Which brings us to our three hypotheses about the success of *A New Hope*. Unfortunately, I'm not going to be quite able to choose among them, but there's something to say in favor of each.

The Greatest Film Ever Seen?

Let's begin with the issue of quality. After its release, *A New Hope* was, of course, immediately recognized as something special. No explanation of its success can disregard that fact. In fact a few people liked it in advance. We have seen that Fox executives were ambivalent or negative, but one of them, Gareth Wigan, cried during a limited screening and concluded that *A New Hope* was "the greatest film [he'd] ever seen." At an advance screening a few weeks later, Steven Spielberg immediately labeled it "the greatest movie ever made."

Audience interest exploded early on, which suggests that people's appreciation of its amazingness, rather than social influences, was the spark. In its first screening for the general public, the audience started cheering at the beginning—and stopped only as the credits rolled. At the Coronet, which Lippincott had worked so hard to book, lines circled around the block. Its manager described the scene as follows: "Old people, young people, children, Hare Krishna groups. They bring cards to play in line. We have checker players, we have chess players; people with paint and sequins on their faces. Fruit eaters like I've never seen before, people loaded on grass and LSD."

At the Avco Theater in Los Angeles, the manager reported that he had to turn away five thousand people over Memorial Day weekend. And before would-be attendees could even begin to navigate the lines, they often had to contend with standstill traffic around urban theaters that effectively shut down driving as a way of getting to a show.

In general, the initial reviews were exceedingly positive, and in some cases they were rapturous. Influential *New York Times* film critic Vincent Canby labeled the film "the most elaborate, most expensive, most beautiful movie serial ever made." A spectacular review in the *San Francisco Chronicle* described it as "the most visually awesome" work since *2001: A Space Odyssey*, while also praising it as "intriguingly human in its scope and boundaries." Joseph Gelmis of *Newsday* went further still, crowning *Star Wars* as "one of the greatest adventure movies ever made" and a "masterpiece of entertainment."

Popular magazines ran stories not only on the movie but also on the phenomenon. "Every TV show news program had done a segment on the crowds waiting to see this amazing movie." At that year's Academy Awards, *Star Wars* was nominated in no fewer than ten categories, including Best Picture. It came away with seven victories. Decades later, numerous directors recall seeing the film and being (to use the technical term) blown away.

Ridley Scott said that he felt "so inspired [he] wanted to shoot [him]self." Peter Jackson said that "going to see *Star Wars* was one of the most exciting experiences that I ever had in my life." Saul Zaentz—a distinguished film producer who would go on to win three Academy Awards—may have been the most moved. Taking out a page in *Variety*, he wrote an open letter to Lucas and his team, congratulating them on "giv[ing] birth to a perfect film[.] [T]he whole world will rejoice with you."

For a generation, Jonathan Lethem captures the feeling this way:

> In the summer of 1987 I saw *Star Wars*—the original, which is all I want to discuss here—twenty-one times. . . . But what actually occurred within the secret brackets of

that experience? What emotions lurk inside that ludicrous temple of hours? *What the fuck was I thinking?* . . . I was *always already* a *Star Wars* fanatic.

It's hard to top the great Lethem, but Todd Hanson does just that:

It was as plain as day, a truism that didn't need to be justified, an axiomatic *fact of nature*, that *Star Wars* was better than anything else you'd previously encountered. It was just *obvious*, kids didn't even need to say it to each other; it was just Known, it was Understood. And not just better, but way better: ten, twenty times cooler than whatever the last coolest thing we'd ever seen had been. . . . It dwarfed whatever it was it had put into second place—you couldn't even *see* second place. Second place was somewhere off the bottom of the page.

So maybe the movie was bound to succeed after all. Recall the "independent judgment" condition of the Music Lab experiment, in which people made their decisions without reference to the views of anyone else. If people saw movies in isolation, and did not learn about what others think or read reviews, there's a good chance that *A New Hope* would still have been a huge hit.

True, we'd have to ask: under those circumstances, how would people have been able to know about the movie? But reasonable people could argue that *A New Hope* is a lot like the very top songs in the Music Lab experiment, in the sense that whatever happened in the early stages, it was going to break out. It was simply too original, too cool, too amazing.

An Exclusive Club?

Possibly so, but let's consider the second hypothesis. Some people, including social theorist Duncan Watts (coauthor of the Music Lab papers), think that essentially nothing is destined to succeed. Even the greatest work needs to benefit from social influences. Yes, absolutely, that includes Shakespeare and da Vinci, too.

For *A New Hope,* there was an informational cascade, big-time, and a reputational cascade too, and network effects helped a lot. The media can spur such cascades, and they definitely did that for George Lucas. On the very day that the movie premiered, the *Washington Post's* review predicted it would be an "overwhelmingly popular" success that could "approach the phenomenal popularity of *Jaws,*" by some measures the most successful movie ever. Just five days after its release, *Time* magazine labeled it "the year's best movie."

A point for the second hypothesis: Its success was self-perpetuating. From the opening weekend, stories about *Star Wars'* popularity, and the wild lines that it attracted, ran in news outlets throughout the country. In June, *Variety* ran an article exploring how telephone operators had become overwhelmed by requests for the telephone numbers of theaters screening *Star Wars.* These operators, reported *Variety,* were forced to memorize theaters' numbers when they found themselves handling a hundred calls an hour.

There were even literal network effects. CBS News anchor Walter Cronkite—the most trusted man in America, the nation's voice—did not ordinarily focus on movies, much less on ones that dealt with the Kessel Run and Jedi Knights. But he devoted time to *Star Wars* in the early weeks of the summer. Just as in the Music Lab, initial popularity spurred additional interest.

According to J. W. Rinzler, the closest thing to the series' official biographer, the gigantic lines that continued to form for *Star Wars* throughout the summer were "fueled to a great degree by person-to-person communication." In an intriguing, brisk analysis, Chris Taylor writes that while "word of mouth in the science fiction community" drew the week-one fans and "[g]lowing reviews" produced viewers for weeks two and three, "[n]ews stories about the size of the crowds brought in the post–Memorial Day crowd." That's a classic description of a cascade.

As Taylor puts it, *Star Wars* was "more than the sum of its box office. It was famous for being famous." He catalogs the early network effects, on which a whole book could easily be written. People who saw the film "were familiar with funny-sounding names and catch-phrases"; they "had joined an exclusive club that knew about 'the Force,' even as everyone had a different theory on what it actually was." Stephen Colbert reported that after seeing *Star Wars*, he and his friends returned to school aware that "everything was different now." Ann Friedman once more: "It offers fragmented audiences a chance to remember what it feels like to be a part of something big that crosses cultural and generational lines. . . . It's nice to leave your niche and experience the truly universal once in a while."

The Perfect Movie for the Time?

But did *Star Wars* also connect with the zeitgeist? Did it have a special resonance for its particular time? Did Lucas, deliberately or by happenstance, end up producing what the public most wanted at that time?

A lot of people think so. In one view, the movie came at a time when the American public, traumatized by a series of demoralizing events, had an acute need for some kind of uplifting mythology. Film critic A. O. Scott captures a widely held view in insisting that the movie's success "represented what looks like the inevitable product of demographic and social forces." Taylor likewise notes that, on the day of *Star Wars'* release, the Dow was at its lowest level in sixteen months, Nixon was being interviewed by David Frost, and the "fingerprints of the [Vietnam] war were everywhere." For his part, theologian David Wilkinson points to the decline of the national economy, emerging ecological concerns, the fresh memories of Vietnam, the lingering dangers posed by Cold War, Watergate, and the stalling of the space program as creating a climate ripe for *Star Wars'* success.

In the documentary *Star Wars: The Legacy Revealed*, journalist Linda Ellerbee notes that "it was not a hopeful time in America . . . we were cynical, we were disappointed, oil prices were through the roof, [and] our government had let us down." In the words of Newt Gingrich, "the country was desperately groping for real change. *Star Wars* came around and revalidated a core mythology: that there is good and evil, and that evil has to be defeated." Indeed, at a time when the president, Jimmy Carter, was taking to the air to encourage Americans to "make sacrifices" and "live thriftily," Americans might be expected to welcome a fantastic adventure a long time ago in a galaxy far, far away.

But maybe not. The cultural explanation, emphasizing the zeitgeist, might just be a way of grasping at straws. To see why this could be so, consider a little contest:

In light of the unique situation of the United States in late May 1977, A New Hope *was bound to succeed because [fill in the blank here].*

You could point to the economy: the stock market, the inflation rate, the unemployment rate. You could point to the international situation: the Cold War, the Soviet Union, China, or Cuba. You could point to Watergate and its aftermath. You could speak of the civil rights movement. You could say something about technology—about the national enthusiasm and ambivalence about it. In one view, *A New Hope* spoke at once to all of those things, and was bound to succeed for that reason.

None of these explanations can be shown to be wrong. The problem is that none of them can be shown to be right. To see why, try the same test, but substitute December 2015 as the date, and *The Force Awakens* as the movie. It would be easy to fill in the blank with reference to the aftermath of the Great Recession of 2008, the rise of the Islamic State, new concerns about technology, or political polarization. People needed a lift! And *The Force Awakens* certainly provided one. But is this explanation right, or just a kind of story, even a fairy tale?

To see the problem, suppose that *A New Hope,* or something quite a bit like it, and with appropriate adjustments for the state of filmmaking at the time, had been released in 1957, 1967, 1987, 1997, 2007, 2017, or 2027. Would it have been a hit or a dud? I say that it would have been a hit. If so, smart people could have done really well on this essay contest: *In light of the unique situation of the United States in late May [fill in the year],* A New Hope *was bound to succeed because [fill in the blank here].* Whatever the zeitgeist—at least within reason—*A New Hope* could easily turn out to be a smashing success.

The upshot: Whenever we say that a product succeeded because of its excellent timing, we might be right, but we might just be telling a tale, not explaining anything. The risk of such timing-was-perfect explanations is heightened for books, music, and movies, where we don't have randomized controlled trials, and where it's easy to say that success was because of an economic downturn, or an economic upturn, or a civil rights protest, or a terrorist attack. Easy—but right?

PUNCH LINES

We've covered a lot of territory, so let's recap.

Some cultural products are the real-world equivalents of the winners in the Music Lab experiment. After the fact, we say that their success was inevitable, because they're great, and because everyone, or a lot of people, think that they're great. But they need early help. Without it, they would look a lot like Sixto Rodriguez or the early Robert Galbraith. *A New Hope* got that early help. Soon after its release, it was famous for being famous, and people wanted to see it because everyone else seemed to be seeing it. Since 1977, that's been its good fortune. *The Force Awakens* benefited greatly from network effects. In a balkanized world, people see it because they don't want to be left out. Star Wars is a bit like the *Mona Lisa*—really famous, and more than good, but the beneficiary of a cultural norm ("this, you have to see") that was far from inevitable.

Some products succeed because they arrive at exactly the right cultural moment. Bob Dylan is terrific, in my view a genius, but his talents and tastes were distinctly adapted for the early 1960s. "Blowin' in the Wind," "A Hard Rain's A-Gonna Fall," and "Like a Rolling Stone"— all these fit well with the time when they were released. They would probably have seemed ugly, or impossibly confusing, in the early 1940s or 1950s, and in the 1970s or 1980s they might have seemed naïve (in the case of the first two songs) or passé (in the case of the third).

True, Dylan is a genius, and also a chameleon, even a shapeshifter. For that reason he might well have figured out something great even if he had been born decades earlier or later. But even though Dylan is a genius, he needed a lot of luck, and big-time network effects, to make it big when he did, even with the cultural resonance. And there is no doubt that his distinctive fit with his time was indispensable to the success of the particular Bob Dylan that actually made it.

But maybe *A New Hope* is just too dazzling to make it necessary, or all that helpful, to speak of some special connection with the culture of the time. Recall the opening scene, with the Imperial Star Destroyer, which seems to be impossibly large. It looks real. You see it from below. Audiences cheered, spontaneously. By the time the movie ends, they have a lot more to cheer about. It hits them where they live. They're still cheering.

Sure, it benefited from cascade and network effects. Sure, it resonated with the culture of the late 1970s. But it was bound to break out. It's too good.

THIRTEEN WAYS OF LOOKING AT STAR WARS

Of Christianity, Oedipus, Politics, Economics, and Darth Jar Jar

There's been an awakening. Have you felt it?

—SNOKE

Unlike a political platform or a religious tract, Star Wars doesn't tell you what to think. It invites speculation. You can understand it in different, even contradictory ways. Sure, the Force is an energy field. (Doesn't everybody know that?) But is it God or at least something spiritual? Do human beings create it? Or is it part of nature? What, exactly, is the relationship between the Light Side and the Dark? What does it mean to "restore balance" to the Force?

The saga is not exactly opaque, like Stanley Kubrick's (insufferable! pseudo-profound!) *2001: A Space Odyssey*. But any Hero's Journey can bear multiple meanings. That's one of the best features of Star

Wars. If it were more didactic or closed, it would be far less interesting, and far less likely to resonate in the way that it has.

For open-textured works, acts of interpretation—*including acts by those who are continuing the texts they themselves started*—have a creative feature. They aren't just about excavation. They involve choice. True, any interpretation has to fit the material. You can't easily say that Star Wars is really about the evils of deficit spending, the problem of climate change, or the importance of increasing the minimum wage. But interpreters have a lot of room to understand it in a way that fits with their own deepest concerns.

With apologies and a salute to Wallace Stevens, here are thirteen ways of looking at Star Wars. Most of them have plausible sources in the movies. A few are nuts, but they're still smart, and in some ways the best of all.

1. CHRISTIANITY

The series, you might insist, is not really the tragedy of Darth Vader; it is an essentially Christian tale about sacrifice, love, and redemption. After all, Anakin Skywalker is the product of a virgin birth. He has no human father. He turns out to be a Christlike figure, dying for humanity's sins, which he incarnates and symbolizes. Drawing on Campbell's monomyth, Lucas produced a highly imaginative reconstruction of Jesus's life, in which the Jesus figure is the sinner, himself unable to resist Satan until the very end, when he sacrifices everything for his child (and symbolically for all children).

Recall that it is the promise of immortality (for his loved ones) that turns out to be Satan's apple. That's how the serpent seduces Anakin, convincing him to give up his very soul. (So there's a Faustian bargain here as well.) But in sacrificing his own life, Anakin defeats the great tempter—and gets his soul back in the process. Loving his son, and killing Satan, he restores peace on earth. (So it's no accident that the word *peace* appears in the crawl in both *A New Hope* and *The Force Awakens*. And Christ is of course the Redeemer.)

That's an essentially Christian tale. "But now faith, hope, love, abide these three: The greatest of these is love" (1 Corinthians 13:13).

Or maybe Luke is the real Christ figure: the Son. Having spent his young years in the equivalent of the wilderness (the farm), he ultimately sacrifices his own autonomy, in a way his life, for the sake of humanity. Perhaps Star Wars has its own kind of Holy Trinity: Anakin, Padmé, and Luke. Star Wars offers a provocative retelling of the biblical story, in which Luke is unquestionably like Jesus, but successfully avoids any kind of crucifixion—and in which the Father is the one who dies twice (in his battle with Obi-Wan and then again with Luke and the Emperor) and rises twice as well (first in his armor, fallen, and finally as a repentant sinner, saved).

The theological resonances in the series are unmistakable, and Christianity looms over the series; it seems to be woven into its very fabric. The whole story is about freedom of choice, good news, and redemption. Is it any wonder that you can readily find books with such titles as *The Gospel According to Star Wars*, *Star Wars Jesus*, and *Finding God in a Galaxy Far, Far Away*?

2. OEDIPUS JEDI

But maybe it is not that at all. Perhaps Star Wars is best understood as something very different, a deeply Oedipal story about fathers, sons, and unavailable mothers. Freud is the right resource, not the Bible.

Maybe there's a complicated sexual undercurrent, and maybe Star Wars is about several kinds of yearning.

Fatherless Anakin is in desperate search for some kind of strong paternal figure, about whom he is inevitably ambivalent. First it is Qui-Gon, then it is Obi-Wan, and finally the Emperor. Anakin, the symbolic son, turns out to be responsible for the death of the third and indirectly the first—and he tries mightily to kill the second. He falls for Padmé, who is a lot older than he is and unquestionably a maternal figure. "You're a funny little boy," she says on first meeting him. "Annie, you'll always be that little boy I met on Tatooine," she says after a long absence, when he's all grown. Isn't that exactly how mothers think about their boys? (And she falls for him!)

Anakin's path to the Dark Side begins only when his mother is killed. In some sense, he's in love with her. Aren't all sons in love with their mothers? On this view, the Tragedy of Darth Vader is a complex and psychologically acute (if somewhat disturbing) reworking of Sophocles's tale.

Luke's story is equally easy to understand in Oedipal terms. He has no father or mother. His youth is a highly ambivalent search for both. He has to make a choice among the various candidates for fatherhood, who are often in fierce competition for his filial devotion: Owen, Obi-Wan, Yoda, Vader, and the Emperor. Remarkably, Luke can be counted as responsible for *all but one of their deaths*. (Old Yoda is the exception, and even that one can be taken as ambiguous.) He's Oedipus—but this particular Oedipus loses his mother because his father, in a sense, killed her. That's its own kind of tragedy.

The all-important qualification is that this Oedipus loves, and redeems, his Dark Father—but the redemption comes only after trying to kill him, and very nearly doing so. Isn't that interesting? Couldn't a good Freudian have a field day with that? On this account, the reworking of *Oedipus Rex* is that the son's love for his father triumphs over his rage. Forgiveness turns out to be all-conquering.

Whether Anakin or Luke is seen as the tale's Oedipus, there is no question that the Freudian echoes in Star Wars help to account for its appeal. It may be Flash Gordon, but the psychological undercurrents are pretty complicated. And of course, Kylo Ren does end up killing his own father, horrifyingly, thus making the Oedipal theme unmistakably clear in *The Force Awakens*.

3. FEMINISM

From the feminist point of view, is Star Wars awful and kind of embarrassing, or actually terrific and inspiring? No one can doubt that *The Force Awakens* strikes a strong blow for sex equality: Rey is the unambiguous hero (the new Luke!), and she gets to kick some Dark Side ass. (Just look at the expression on her face when she has a go at Kylo.) There's also General Leia, and Captain Phasma, and women can be found in multiple positions of leadership.

By contrast, the original trilogy and the prequels are easily taken as male fantasies about both men and women. The tough guys? The guys. When you feel the Force, you get stronger, and you get to choke people, and you can shoot or kill them, preferably with a lightsaber (which looks, well, more than a little phallic, and the longer, the better). And for men of a certain age, the most memorable scene in the whole series can be found in *Return of the Jedi*, when Leia is tied up and in a bikini. Isn't that more than a bit retrograde, or worse?

But there's another view. Is everything redeemed, because Leia gets to strangle her captor, using the very chain with which he bound her? Is that the real redemption scene in the series?

In the first two trilogies, Princess Leia and Padmé Amidala are of course major figures. They can easily be seen as central characters, the best clues to the deepest meaning of the series. Created in the 1970s, Leia was way ahead of her time. Sure, Obi-Wan Kenobi was her only hope, but she's no damsel in distress. On the contrary, she's a military leader, the most important of them all, and the person who sets the whole rebellion in motion. For many viewers, she's a feminist icon. A natural commander, she is the smartest, the wisest, the most steadfast, and the bravest. She knows how to shoot, and she doesn't hesitate to do exactly that. Most of the time, she's the boss. It's unsurprising that in *The Force Awakens*, she's a general. We would have to stretch to see the first trilogy as feminist, but there's a good argument that for its time, it was fine and even inspiring from the feminist point of view—and a lot of women were inspired by what Lucas did.

In the prequels, the much-maligned Padmé is solid as a rock. She's also the person who sees things most clearly (even if some of her lines aren't so great). She's a leader, too—a queen and then a senator. She has more than a glimpse, early on, of what is happening to the Republic. By contrast, the men—Anakin, Luke, Han—are pretty clueless.

However we might evaluate the first trilogies, it remains true that the force that really awakens, in Episode VII, is sex equality. Rey is the strongest character and the best—the most interesting, the funniest, the sharpest, the most complicated, the most Force-sensitive. (And so the public outcry was entirely justified when Rey was originally excluded from games, toys, and other tie-in products.) We're not sure, yet, whose daughter she is, but make no mistake: in the deepest sense, she's a Skywalker.

4. THOMAS JEFFERSON, JEDI KNIGHT

The series could easily be seen as profoundly political, meant to emphasize the need for rebellion or at least for maintaining the potential for one. From the novelization of *A New Hope:* "Only the threat of rebellion keeps many in power from doing certain unmentionable things." There's more than a mild echo here of Thomas Jefferson, who thought that turbulence itself is "productive of good. It prevents the degeneracy of government, and nourishes a general attention to the public affairs. I hold it that a little rebellion now and then is a good thing, and as necessary in the political world as storms in the physical." (Rebellion! Did Lucas read Jefferson?)

On this view, the real topic of the series is the Jeffersonian one, pointing to the value of rebellion and self-government, the virtues of republics, and the vices of empires. The Empire and the First Order despise turbulence, which they see as a form of chaos; they want order, which is, in their version, another word for choicelessness. Lucas saw things at least partly this way, regarding the Emperor as a kind of Richard Nixon figure. (Nixon was Mr. Law and Order.) Lucas even regarded the rebels as the Vietnamese—and the Empire as where the United States was headed (in a decade!). Abrams continues the Jeffersonian theme with the battle between the First Order and the Resistance.

The series has a kind of giddiness, to be sure, but maybe it has a far more serious message, which is about how essential it is to keep a close eye on political leaders, who must be closely monitored by a vigilant public. Actually I think the series is indeed telling us that. It makes a consistent opposition between order and freedom of choice, and there's no question which of these it chooses. Its insistence on choice-making at the individual level (Luke, Han, Anakin, Rey, Finn) is replicated in its politics.

5. ORDER YES, CHAOS NO

But maybe exactly the opposite is true. Maybe the Jedi are the wrongdoers—baffled, stumbling, unable to maintain stability. Maybe Emperor Palpatine is the secret hero, after all. Maybe that's the dark heart of the Star Wars series. Maybe that's where its sympathies really reside.

Does the last argument seem nuts? At least since 2002, intelligent people have argued that it's right. Under the Republic, things were pretty chaotic, and the Jedi Knights failed to ensure order, which human beings really do need. (So maybe Nixon wasn't all wrong.) Part of the tension of the movies stems from the unmistakable appeal of a strong leader, who can unite people and rescue them from chaos. Powerful leaders of all kinds insist that that is what they are doing; Vladimir Putin is just one example, and in 2015 and 2016, Donald Trump's surprising run for the presidency reflected something similar. (Hitler, you might respond. George Washington, they might reply.) The Jedi failed miserably in the endeavor of restoring order; the Emperor succeeded. *The Force Awakens* is all about the tension between order and chaos, and it's secretly in favor of the former. In that respect, it is following the first two trilogies.

In the words of one essayist:

Make no mistake, as emperor, Palpatine is a dictator—but a relatively benign one, like Pinochet. It's a dictatorship people can do business with. They collect taxes and patrol the skies. They try to stop organized crime (in the form of the smuggling rings run by the Hutts). The Empire has virtually no effect on the daily life of the average, law-abiding citizen.

A stronger statement:

It is the Empire, not the Rebel Alliance, that offers the best hope for the future of the race. It is the Empire, not the Rebel Alliance, that is best equipped to bring peace and prosperity to this troubled galaxy. . . . By resisting, subverting, and ultimately destroying the Empire, the rebels have bequeathed their children a chaotic, primitive, technologically retrograde society that will almost certainly collapse into anarchy within a generation.

Clever, maybe, but okay, I agree, the idea that Star Wars favors the Empire is definitely nuts. (*The Force Awakens* does create some complications, but it would be a real stretch to contend that the New Order has things right.)

6. BEHAVIORAL STAR WARS

For decades, behavioral economists and cognitive psychologists have explored how human beings depart from perfect rationality. It's not exactly news to announce that we aren't computers; in deciding what to do, people don't quantify the expected outcomes and run probability calculations. Nor are we irrational, at least most of the time. What behavioral scientists have shown is that human beings suffer from *predictable biases*. The Jedi Masters who have uncovered such biases have won at least five Nobel Prizes in economics. Daniel Kahneman, author of the magnificent *Thinking, Fast and Slow*, is the most famous of them; to many people, he's a real-world Yoda. (A good life lesson from Kahneman: "Nothing in life is as important as you think it is when you are thinking about it." Think about that. It's important.)

Some examples of human foibles: People are overconfident. ("The Resistance has no chance against the First Order; this is its final hour!") We tend to focus on today and tomorrow, not next month or next year ("present bias"). We display unrealistic optimism. (About 90 percent of drivers have been found to believe that they are better than the average driver. Or: "Everything is happening as I have foreseen.") We suffer from inertia and so we procrastinate. Instead of examining statistics, we use simple heuristics, or rules of thumb, in assessing risks. (Did a crime occur in my neighborhood in the recent past? Did the Empire attack a planet just like mine?) Our judgments are systematically self-serving. ("What's fair is what's best for me!") We dislike losses far more than we like equivalent gains ("loss aversion"). So it shouldn't be surprising that golfers do better putting for par than for birdie (a bogey is a loss, and people hate losses), or that if you want people to conserve energy, you'd do best to emphasize that they would lose money if they fail to use energy conservation techniques, rather than that they would gain money if they use such techniques.

In fact the modern era of behavioral science started in the late

1970s—just around the time that *A New Hope* was released. Can that possibly be a coincidence?

It can't be! Star Wars is a series of case studies in behavioral biases. Darth Vader and Emperor Palpatine suffer from both unrealistic optimism and self-serving bias; they think that everything is going to work out in their favor. Their overconfidence leads them to make big mistakes at critical moments. (Snoke has the same problem.) But Star Wars knows that behavioral biases are not limited to those who favor the Dark Side. One of our heroes, Han Solo, is also subject to optimistic bias:

> **C-3PO:** Sir, the possibility of successfully navigating an asteroid field is approximately 3,720 to 1!
> **HAN:** Never tell me the odds!

Of course things work out well for Han, certainly in navigating that asteroid field, as they do not for Vader and Palpatine. Optimistic bias can help you in the roughest times. (But in *The Force Awakens*, Han's characteristic tendency to unrealistic optimism created a very big problem. He should have had a bad feeling about that.)

Both Luke and Rey suffer from inertia and its close cousin, "status quo bias," which refers to people's tendency to prefer things to stay as they are, even if it's really a good idea to make a change. Inertia and status quo bias are why Luke chooses to decline Obi-Wan's suggestion that he accompany him to Alderaan. They also account for Rey's refusal of Luke's lightsaber. The good news is that the Force runs strong in their family, and so they're able to overcome their behavioral biases. (Isn't that what the Force is for, after all? Isn't that what Star Wars is telling us?)

It would be easy to teach a whole course on behavioral economics with close reference to Star Wars. Maybe that's how the series is best understood. (But you'd have to be biased to think so.)

7. TECHNOLOGY

Maybe the series is a cautionary tale about the dehumanizing effects of technology. Lucas certainly saw it that way. He was obsessed with that topic—and with what technology does to us.

A New Hope begins with droids. In a sense, they are the narrators of the tale, and they have human characteristics; that's part of their charm. BB-8 plays the same role in *The Force Awakens* as R2-D2 plays in *A New Hope;* the two are like cute pets or loyal younger siblings. (It would have been interesting to make them potentially disturbing, but Star Wars doesn't go there. BB-8 gone rogue? That could be a bit scary.) But dehumanization through machines, and machine parts, plays a large role throughout the series.

In 1962, Lucas himself had a near-fatal motorcycle accident. In his words: "In high school I lived to be a race car driver and I was in a very bad accident. . . . I was hit broadside by a car that was going about 90 miles per hour. . . . I should have been dead." Machines helped to keep him alive. Whether or not his own experience is responsible for his focus on the symbiotic relationship between human beings and machines, there is no question that Star Wars focuses on that relationship.

Darth Vader is frightening because he is part person, part machine. Obi-Wan to Luke in *Return of the Jedi:* "When your father clawed his way out of that fiery pool, the change had been burned into him forever—he was Darth Vader, without a trace of Anakin Skywalker. Irredeemably dark. Scarred. Kept alive only by machinery and his own black will." That is a fact, but it is also a symbol: falling to the Dark Side, he loses much of his humanity—a prescient warning for those who live in an age of machines. (Check your email lately?) It's no wonder that a farm boy, from an isolated land, is the one who restores peace and justice to the galaxy.

8. JEDI JIHAD

Were the Jedi engaged in some kind of Jihad? Are the Rebels terrorists? Maybe the original trilogy is all about the radicalization of Luke Skywalker. Maybe it's a case study in how radicalization works. (Recall the force of echo chambers.)

Luke begins as that innocent farm boy, with no particular religious convictions. He is isolated and rootless—an excellent target for extremists. Sure enough, he embarks on what an online commentator describes as a "dark journey into religious fundamentalism and extremism." A disaffected and somewhat lost young man, in search of something, he comes across Obi-Wan Kenobi, plainly a religious fanatic, who follows self-evidently extremist ideas about the Force. "Within moments of meeting Luke, Obi-Wan tells Luke he must abandon his family and join him, going so far as telling a shocking lie that the Empire killed Luke's father, hoping to inspire Luke to a life of jihad."

Obi-Wan succeeds in that endeavor. Gradually he convinces Luke to believe in his radical cause, in the process convincing him to accept cultlike religious convictions. To complete Luke's radicalization, he "says a Jedi prayer while committing suicide. Can you think of any other groups who try to inspire terrorism by yelling a prayer before a suicide attack?" In the end, Luke becomes a full-blown terrorist.

Okay, that one's nuts, too.

9. THE DARK SIDE AND THE DEVIL'S PARTY

Say it loud and say it proud: Vader steals the show. Who's the most memorable character in the series? Vader is the most memorable character in the series. No one else comes close.

A New Hope, The Empire Strikes Back, and *Return of the Jedi* are most captivating when he's on-screen. They tend to sag a bit when he's not (at least *Return of the Jedi* does). My son Declan wore a Darth Vader outfit this past Halloween, and a lot of six-year-olds make the same choice. How many Luke Skywalker costumes are there, anyway? And have you seen an Obi-Wan costume? Ever?

Sure, Luke is appealing, and he's really nice, and he even becomes a Jedi. But does any boy ever watch Star Wars and think: I want to be Luke! He's too earnest for that. Besides, he doesn't get the girl. She's his sister!

Han Solo is a lot cooler, and he's my personal favorite. He's a rogue, but he's also retrograde, and that's a bit of a problem. He was a 1950s person, even in the 1970s, and he's a 1970s person now. Harrison Ford does great by him, but still, is it unfair to wonder whether he's a nerd's conception of a cool person?

By contrast, Vader doesn't get dated. He's sleek, and he's big, and he can choke people just because of his thoughts. Also, he doesn't give a shit. No wonder Kylo Ren idolizes him. (And after the release of *The Force Awakens*, there were plenty of Kylo toys. Declan got one for Christmas.)

In a brilliant essay, Lydia Millet writes that Vader "was the most erotic figure in the Star Wars family, and the only tragic one, and because of this he had a terrible beauty." An aristocrat, "he had poise, elegance and good manners." He was also "the only question Star Wars posed to its audience, the only mystery presented." Mastery, distance, and command are his defining features. In Millet's view, he "has an erotic charge because he gets what he wants." (True.) He makes the Dark Side seem sexy. (It is, isn't it?)

The great William Blake, writing about *Paradise Lost*, one of the most religious texts in the English language, pronounced: "The reason Milton wrote in fetters when he wrote of Angels and God, and at liberty when of Devils and Hell, is because he was a true Poet and of the Devil's party without knowing it." Blake had a point about Milton, who fell for Satan's energy and charisma. Milton was indeed a true poet, and he had a conception of liberty, and so he visited, with relish, the Dark Side.

By the way, Blake himself was also a true poet, and the Dark Side was very much with him. "Sooner murder an infant in his cradle than nurse unacted desires." And: "Enough! Or Too much." And: "The tigers of wrath are wiser than the horses of instruction." And: "Those who restrain their desires, do so because theirs is weak enough to be restrained." And of particular relevance to Star Wars: "Without contraries is no progression. Attraction and repulsion, reason and energy, love and hate, are necessary to human existence."

Was George Lucas also of the Devil's Party? Not really. In the end, he's a good guy; he's Luke. But he did get tempted. Lucas wrote *A New Hope* about Luke (his namesake), but it was the character of Vader who captured his imagination, and so the Sith Lord took over the narrative.

There is no question that to some people, the Emperor himself has a satanic appeal; he's a seducer, and a really good one. He's full of a kind of lust. That gives the movies their necessary tension. In the best of them all—*The Empire Strikes Back*, of course—Millet notes that "the forces of good are routed and evil enjoys unquestioned triumph."

She's right to insist that "Darth Vader is the fulcrum, the focal point, the emotional center of the Star Wars saga."

Like Blake and Milton, Lucas knew well the attraction of the Dark Side. He went there. As he noted, "People like villains because they're powerful and they don't worry about the rules." But there's something even more primal in their appeal. Palpatine to Anakin, with an erotic charge: "Good, I can feel your anger. It gives you focus . . . makes you stronger." Palpatine to Luke: "Good, I can feel your anger." And from the novelization, while Luke is in ecstatic battle with Darth Vader: "And in this bleak and livid moment, the Dark Side was much with him."

STAR WARS VS. STAR TREK: AN ASIDE

Who's better, Michael Jordan or LeBron James? (Jordan, because he'd rip your heart out.) Abraham Lincoln or Franklin Delano Roosevelt? (FDR, because he saved the country twice, and because he was cheerful rather than melancholy, and so more characteristically American.) Meryl Streep or Julianne Moore? (Close, really close, but Moore, because *you never think that she's acting.*) The Beatles or the Rolling Stones? (The Stones, because they know the Dark Side.) Immanuel Kant or John Stuart Mill? (The gentle, clear-headed Mill, eight days a week.) Taylor Swift or Adele? (Swift, by a country mile, because her sense of mischief and fun ensures that she is never, ever saccharine.) Ronald Reagan or Barack Obama? (Obama, but you knew I'd say that.)

Star Wars or Star Trek?

There's a lot to be said for Gene Roddenberry's masterpiece. Consider an early episode, *The Enemy Within*, which offers its own depiction of the contest between the Dark and Light Sides. By a transporter malfunction, Captain James Tiberius Kirk is converted into two people: one is good and the other is evil. The evil Kirk is aggressive, even violent; he's angry, cruel, and selfish. He wants what he wants when he wants it. He's out of control.

You might well think that the good Kirk is the real one and that the bad one isn't. In its early scenes, that's exactly the reaction that the episode invites. But you'd be wrong. The good Kirk, speaking of his counterpart: "He's like an animal. A thoughtless, brutal animal. And yet it's me. Me!"

What emerges is that the two are equally indispensable to the Kirkness of Kirk—that without the apparently evil side, Captain Kirk is indecisive, immobilized, passive, weak, pale, a kind of ghost. McCoy to the good Kirk: "We all have our darker side. We need it! It's half of what we are. It's not really ugly. It's Human." That's true, and in a way, it's subtler and better than anything on the topic in Star Wars. Also good, from *Frame of Mind*, an episode of *Star Trek: The Next Generation*: "Sometimes it's healthy to explore the darker sides of the psyche. Jung called it 'owning your own shadow.' . . . Don't be afraid of your darker side. Have fun with it."

The original series is more lovable, but my vote for the best Star Trek episode, ever, goes to *The Inner Light*, again from *The Next Generation*. Captain Jean-Luc Picard is transported to the planet Kataan, where his wife convinces him that his memories of having been a starship captain are a kind of delusion, produced by an illness. His real name is Kamin, and he has a beloved wife and comes to have two beloved children, a daughter and a son. (He tells his daughter: "Live now. Make now always the most precious time. Now will never come again.") On Kataan, he grows older, and then old, and he has a grandson. His life is full and soft and good. But eventually he learns that because of increasing radiation from the sun, his entire world is doomed and will soon be destroyed.

Aware of that tragic fact, Kataan's leaders placed memories of their culture into a probe and launched it into space. They hoped, desperately, that the probe would find someone who could learn about their species and ensure that it would not be forgotten. In his last moments as Kamin, having lived his life for what seemed to be decades, Picard understands. He's heartbroken, and stunned: "Oh, it's me, isn't it? I'm the someone . . . I'm the one it finds."

What makes this episode so beautiful is that it captures both the preciousness and the impermanence of our eras, our cultures, and our individual lives. (Were the 1960s Kataan? Were the 1990s? What about this decade?) In a short period, Picard/Kamin is able to see himself as a relatively young man, as a husband, as a parent, as a grandparent, as elderly and near death. In a sense, all of his selves become present to him. This is all the more moving because we are speaking of a civilization that is now entirely lost. *The Inner Light* definitely doesn't move like a son of a bitch—but it gets close to the center of the human heart. Star Wars does that too, and we'll see why, but nothing in it quite compares to *The Inner Light*.

In terms of the visuals, Star Wars is infinitely better. It's far more exhilarating; it produces a feeling of whoa and OMG, as Star Trek's visuals never do. Star Wars has a continuing sense of mystery. Unlike Star Trek, it makes you try to connect dots. It's cooler, and it's more awesome. Star Trek is far more literary, and in fact, many of its greatest episodes were written by novelists. Even more than Star Wars, it makes you think about enduring questions. Star Wars is not unlike a series of paintings; Star Trek is closer to a set of novellas.

At their best, they're both great. Which is better?

Philosophers speak of the idea of "incommensurability." What they mean, roughly, is that we value qualitatively things along different metrics, and so they can't really be ranked. Sure, $1,000 is better than $500. But what's better, a beautiful mountain, a terrific athletic performance, dinner with a great friend at a great restaurant, an amazing song, or a specified amount of money? One answer is that all of these are valued in different ways. You might try to rank them if you wish, but you might want to keep in mind the qualitative differences among them. Star Wars and Star Trek are good in different ways, and in fairness, you can't really rank them.

But Star Wars is better.

"YOUR FOCUS DETERMINES YOUR REALITY"

Interpretation often involves an effort to show that *everything fits together*, if only we look hard enough. Interpreters try to find scripts. Their efforts usually work by assembling telling details, some of them seemingly irrelevant. An inflection of voice, a wry smile, an inappropriate laugh, a placement of a comma, a choice of the word *the* rather than *a*, an addition of an *s* at the end of a word—all these can suggest a plan where we might otherwise be at sea. Some examples:

- Why was the head of the Department of Justice, the attorney general of the United States, out of the country *on the very day of a racially motivated murder in Atlanta?*
- Why, exactly, did Obi-Wan Kenobi smile immediately before he was struck down by Darth Vader? Is it because he wanted to die? (Maybe!) Because he had gone over to the Dark Side? (Nah.)

- In *A New Hope*, C-3PO and R2, carrying the plans for the Death Star, are ejected from Leia's ship in an escape pod—which lands suspiciously near the place where Luke, Anakin, C-3PO, and R2 have all lived in at some point. Could that possibly be a coincidence? *What are the odds?*

Conspiracy theorists are masters of this approach, and they love such questions. They insist on finding scripts, full of hidden clues. ("Everything is unfolding according to plan." "The truth is out there." Bush's endgame, Obama's endgame, Putin's endgame, the Pope's endgame.) Even more than the rest of us, they overlook the extent to which random or arbitrary factors are responsible for what happens. Whether or not they're crazy, they're certainly not dumb, and they are hardly ignorant. On the contrary, they tend to be specialists. They know a ton. They comb through vast materials, finding countless patterns and links (aha!), and then declaring their suspicions to be confirmed. Don't bother arguing with them; they know a lot more than you do. Don't even try.

Take a look at any relevant writing about the assassination of John F. Kennedy, or about the attacks of 9/11, or about what's going to happen in coming Episodes, and you'll see the point. When such writing is done well, it shows a mastery of a uniquely human ability: to connect seemingly random dots.

10. HAN THE PADAWAN

Star Wars fans are themselves experts at this kind of thing. Here's a good one: Han Solo is able to use the Force—but he doesn't know it.

How else could he escape all those bounty hunters? Without the Force, how could he possibly have avoided that point-blank shot by Greedo, simply by craning his head slightly to the right? (I am bracketing the admittedly important question of whether Han or Greedo shot first.)

And when Han says, "Kid, I've flown from one side of this galaxy to the other," and adds, "There's no mystical energy field that controls my destiny," why does Obi-Wan give that knowing smile? Isn't it because Obi-Wan is letting us in on a little secret: Han Solo is himself a kind of Padawan! Maybe that casts a new light on what happens between Han and Kylo in *The Force Awakens*? Maybe what we saw there is not quite what we think we saw there?

11. ANOTHER BROTHER!

In 2015, people started speculating that Luke and Leia have a brother! The supposed clue lies in the crawl for *The Force Awakens*, which says that Leia "is desperate to find her brother Luke and gain his help in restoring peace and justice to the galaxy." That's a clue because the name "Luke" is not surrounded by commas, as it should be.

Hence the theory: "His name is unnecessary information, as Leia has only one brother, meaning it should be offset by commas." Without those commas, "her brother Luke" seems to suggest that there is another brother, as in, "her brother Snoke" or "her brother Boba."

Or not.

12. BUDDHIST STAR WARS

Take a look at Yoda, there in *The Empire Strikes Back*, sitting in his cape. He looks a lot like Buddha, doesn't he?

Are the Jedi Buddhists? They certainly emphasize the importance of detachment—of transcending fear and hatred through a form of serenity. Yoda's famous words: "Fear Leads to Anger / Anger Leads to Hate / Hate Leads to Suffering." Compare Buddha's version: "There is Suffering / There is a cause of Suffering / There can be an end to Suffering / The eightfold path brings about the end of Suffering." That eightfold path can easily be seen as the model for Yoda's training of Luke. (It begins with a vision of the nature of reality and the path of transformation.)

The Jedi order looks a lot like Buddhist orders, and the master-Padawan relationship seems to mirror the teacher-student relationships in Buddhism. In those relationships, a great emphasis is placed on the idea of "mindfulness," or being in the present moment, rather than the past or the future. You can read *A New Hope*, and Obi-Wan's teachings, as being all about mindfulness. In the crucial moment, when Luke uses the Force to destroy the Death Star, it's all about being present.

But we don't even have to speculate. In *The Phantom Menace*, Qui-Gon's constant advice to Anakin is "be mindful." For example:

OBI-WAN: But Master Yoda says I should be mindful of the future.
QUI-GON JINN: But not at the expense of the moment.

And Qui-Gon also says this: "Remember, concentrate on the moment. Feel, don't think. Trust your instincts." So Star Wars is not about Christianity at all; it's Buddhist. Is there any wonder that an ordained member of Thich Nhat Hanh's Buddhist community, Matthew Bortolin, has written a whole book on the topic, called *The Dharma of Star Wars*?

13. DARTH JAR JAR

Here's an especially bold, and bizarrely credible, exercise in dot connection, which went viral in late 2015: *Jar Jar Binks is a Sith Lord.* ("Search your feelings, you know it to be true.") According to Reddit user Lumpawaroo, Jar Jar "was not, as many people assume, just an unwitting political tool manipulated by Palpatine—rather, he and Palpatine were likely in collaboration from the very beginning, and it's entirely possible that Palpatine was a subordinate underling to Binks throughout both trilogies."

In the prequels, Lucas's original plan was to give Jar Jar Binks a prominent role in all three movies, akin to that of Yoda. He acts like a fool, a joker, an idiot, but he knows just what he's doing. Crazy like a fox, he's the brain behind the scenes. He is playing the Jedi like a violin. According to this account, Lucas had to abandon his plan because so many people hated Jar Jar, couldn't stand his presence, and depicted him as a product of racism.

True, all this sounds crazy, but Lumpawaroo elaborated the theory with such care, diligence, energy, and obsessiveness that it actually made a kind of sense. One result was a website dedicated to the idea, darthjarjar.com. Is it right? Lucas had to issue a public denial to prevent this account from getting real momentum.

Now he would deny it, wouldn't he?

THE WORLD IS RUDDERLESS

Alan Moore, the great graphic novelist and author of the sensational *Watchmen*, spent many years studying conspiracy theories. Here's what he ended up concluding:

The main thing that I learned about conspiracy theory, is that conspiracy theorists believe in a conspiracy because that is more comforting. The truth of the world is that it is actually chaotic. The truth is that it is not The Iluminati, or The Jewish Banking Conspiracy, or the Gray Alien Theory. The truth is far more frightening—Nobody is in control. The world is rudderless.

If Lee Harvey Oswald traveled to Moscow on a certain date, or if on 9/11, a lot of Jews happened not to show up for work in New York, we might start to put some pieces together. But it might be best not to try. Details that appear to be telling often tell us nothing.

Psychoanalysts help a lot of people, but they too neglect randomness and serendipity. They purport to find patterns in dreams, thoughts, and behaviors, *even when they are creating those patterns, not finding them*. They think that they can fit the pieces together, and the best ones are terrific at it. They are conspiracy theorists, at least of a sort.

To their credit, they do know that sometimes a cigar is just a cigar. But if you dreamed, or say that you dreamed, that you are a Jedi Knight, and if you happen to have had a horrible encounter with your father the previous week, you should hesitate before insisting that the particular dream reveals some important psychological fact. Sure, you could speculate. You might have wanted to demonstrate that you are more powerful than your father, or to escape your father, or to be a Jedi Knight so you could kill your father. But maybe not.

Literary critics are a lot like conspiracy theorists. Shakespeare used the word *nature* at important points in *King Lear,* and he also used the word *potency* in close proximity; is *Lear* a cautionary tale about the indomitable power of nature in the face of human weakness? If we parse Shakespeare's texts closely enough, might we find coded messages, suggesting highly subversive political messages? Was Shakespeare a rebel? Some people so insist; there is at least one whole book on the topic. For that matter, maybe Shakespeare did not write Shakespeare's plays; perhaps they were written by Francis Bacon, Edward de Vere, or Christopher Marlowe. A close reading of the plays might provide evidence to that effect. It is easy to find many volumes of work in this vein. My personal favorite, by Sir Edwin Durning-Lawrence, is called *Bacon Is Shakespeare.* Almost every chapter ends with three words, in capital letters: BACON IS SHAKESPEARE.

Consider in this light *The Bible Code*, which contends that the Bible is filled with hidden messages, which you can find by looking at (say) every fiftieth letter of the book of Genesis. If you do that, you might be amazed to find the names of famous rabbis, along with their dates of birth and death, and you might also find some important predictions about future events. Maybe the Bible predicted the Holocaust, or the rise of communism, or the attacks of 9/11, or *The Force Awakens.* A lot of people were taken in by those who argued, in 1998, that the Bible really contains such a code. But it's a fraud.

There is an explanation for all this in the domain of visual perception: Our brains are wired to see patterns where they do not in fact exist (a phenomenon called "patternicity" or "apophenia"). What do you see here, on Mars?

(These are three photos of the same object from NASA.)

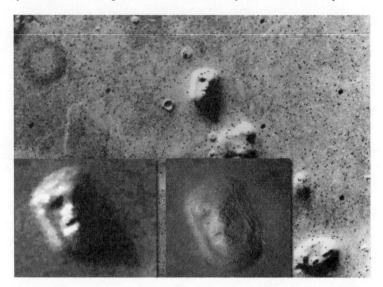

NASA.

Is it a face? Isn't it a Stormtrooper?
Nope. It's just a rock.

FATHERS AND SONS

You Can Be Redeemed, Especially If Your Kid Really Likes You

"Come, now, let us reason together," says the Lord.
"Though your sins are like scarlet, they shall be as white
as snow; though they are red as crimson, they shall be
like wool."

—ISAIAH 1:18

He wanted me to go into his business. I said, "I'm
absolutely not going to do it." He sold office equipment in
a store. I said, "I will never go to work every day doing the
same thing day in and day out."

—GEORGE LUCAS

My son Declan, the Darth Vader fan to whom this book is dedicated, is now six years old. Three years ago, he was joined by his sister Rian. Declan is very close to his mother, and when Rian was born, I had an acute sense that he might feel some kind of threat and loss. Within a few weeks after she came home, I found myself singing my son a song, from the depths of my unconscious, that managed to be both idiotic and offensive. It had just one line, sung over and over, and with some combination of giddy delight and utter certainty: "DADDIES ARE FOR BOYS, AND MOMMIES ARE FOR GIRLS."

Declan pretended to dislike it, or at least to think that it was wrong. His response was to sing right back, to the same tune, "DADDIES ARE FOR GIRLS, AND MOMMIES ARE FOR BOYS." I responded by saying, "Declan, you must be very tired, and all confused. You're singing the song all wrong. Did you get enough sleep last night?" He came right back at me: "Daddy, you must be working too hard, and all confused. You're singing the song wrong. Did you get enough sleep last night?"

Whenever I came back from a business trip, I'd pick him right up and tell him: "Declan, this is completely amazing, you're not going to believe it! I was watching a news program in a hotel room in California, and a commercial came on, and there were these people, and everyone started singing, 'DADDIES ARE FOR BOYS, AND MOMMIES ARE FOR GIRLS.' So it's true!" And he would answer, "Daddy, while you were gone, I was watching a baseball game at home, and a commercial came on, and everyone started singing, 'DADDIES ARE FOR GIRLS, AND MOMMIES ARE FOR BOYS.' So you must be confused. Are you very tired?"

While Declan still pretends to dislike our little song, he has always known what it's really about. The song originally said: "You have a baby sister, and she takes some of your mother's time, but your father is here for you." Its meaning is simpler now: "I'm here for you."

There's some evidence that he gets that. Whenever my wife and I are out late to dinner, he tends to fall asleep on our bed. As I pick him up, to carry him off to his own room, I start whistling the insipid tune to a particular song. He's sound asleep. But every time, he smiles.

Rian is three years old now, and she's heard the song countless times, in its two different versions. Sometimes she sings, with delight, "DADDIES ARE FOR GIRLS, AND MOMMIES ARE FOR BOYS." Sometimes she sings, with equal delight, "DADDIES ARE FOR BOYS, AND MOMMIES ARE FOR GIRLS."

Both versions are right. They're complementary, not contradictory. Rian is a smart girl, and she knows exactly what she's talking about.

REGRET

Ask any adult this question: "What's your greatest regret in life?"

There is a decent chance that the answer will be "I wasn't good enough to my parents." That's an especially likely answer if a parent has passed away. If so, and if you were estranged or just on bad terms, that answer sears the soul. And even if you were a very good son or daughter, you might think that you weren't good enough, and you might well have that same regret. As you voice it, you might tear up a bit (as I'm doing now).

For those who give this answer, imagine what it would mean for a parent to come back from beyond the grave and tell you, "You were the best kid I could possibly imagine having, ever, and you shouldn't have a moment of regret, and I love you." Or if that's unrealistic, how about this: "Sure we struggled. That's what we were supposed to do! No one's perfect, least of all me. We both made mistakes. Welcome to the human race. You're my kid, and we're in each other's hearts, and I love you."

If you ask people that question about their greatest regret, there's also a good chance that you'll hear this answer, at least from people of a certain age: "I wasn't a good enough parent." People with grown children almost never ask, "Why did I spend so little time working?" Instead they think, If I could do it all over again, I would have been a much better parent. If you are estranged from your child, or just on bad terms, that's a searing thought. But even if you were an excellent parent, you might think that you could have been a lot better.

For those who give this answer, what would it mean for a child to drive to your home or to get in an airplane, and to say, "Gosh did you love me, and I felt it every day, and I feel it now. You shouldn't have a moment of regret. Of course you weren't perfect. But to me, you were the best parent in the world." Or if that's too much, how about this? "You weren't the best parent, and I wasn't the best child. But it's never too late. I love you. How about dinner tonight?"

That's one of the things Star Wars is most deeply about—fathers, sons, and redemption. In its own way, it's saying, over and over again, that DADDIES ARE FOR BOYS. It is pointing to the indispensability of paternal love, and it has a lot to say about the lengths to which people, whether boys or girls, will go to get it.

"THIS BIG BREAK"

To every child, boy or girl, a father must seem, at times, to be a kind of Vader—large, tall, frightening, with a booming deep voice, insanely powerful, and at least potentially violent. For any child, boy or girl, a father is both Jedi and Sith—Obi-Wan Kenobi, gentle and calming and good, and Vader, fierce and terrifying. Of course, every father offers his own combination. But almost every one seems to have easy access to the Dark Side, at least to a child, and with his immense power, he appears capable of anything.

In the first trilogy, Lucas was able to get quite primal about fathers and sons, and while his tale speaks to everyone, he's given some personal hints about why. His relationship with his own father, George Sr., was troubled, in some ways even tortured; it was full of disappointment and mandates and prohibitions. As one of Lucas's interviewers noted, George Sr. was known as a "domineering, ultraright-wing businessman," and those who "know Lucas have always insisted that the tortured relationship between Darth and Luke springs, in many ways, from Lucas's relationship with his own father."

Lucas's father did not (so far as we know) try to convince him to go to the Dark Side, or to rule the universe as father and son, but he did urge him to abandon his dreams and to join him in the family business. "My father wanted me to go into the stationery business and run an office equipment store. . . . He was pretty much devastated when I refused to get involved in it." By all accounts, their confrontations on this subject were turbulent, and for a time, they ended up estranged. (It's worth pausing over that. Even if it's temporary, an estrangement between a parent and child is extraordinarily painful. Been there, done that.)

As Lucas put it, matter-of-factly but with defiance, "At eighteen, we had this big break, when he wanted me to go into the business, and I refused." As his father recalled, "I fought him; I didn't want him to go into that damn movie business." Even though George Sr. uttered those words years later, you can feel the fire: "damn movie business."

True, there were no lightsabers. No one lost a hand. But every son yearns for his father's approval, and it was not easy for Lucas to get his.

Movingly, Lucas says, "You only have to accomplish one thing in life, and that's to make your parents proud of you." And every child yearns to know who his parents really are. Do we ever find out? I'm not sure.

Referring to himself and Steven Spielberg, Lucas once noted, "Almost all of our films are about fathers and sons. Whether it's Darth Vader or E.T., I don't think you could look at any of our movies and not find that." That's quite a statement for someone who has done a wide range of movies, and whose most famous ones seem to deal with planets, spaceships, and droids. And more personally, and quite gently: "Parents try as hard as they can to do the right thing. They aren't purposely out to get you. They don't want to be Darth Vader."

Lucas himself was able to reconcile with his father, though it took years for them to come back together. He packs a lot of pain and understanding into these words: "he lived to see me finally go from a worthless, as he would call 'late bloomer' to actually being successful. I gave him the one thing every parent wants: to have your kid be safe and able to take care of himself. That was all he really wanted, and that's what he got."

It's not irrelevant that after *Return of the Jedi*, Lucas abandoned Star Wars, and movie-making, for just one reason: he wanted to be a good father. He retired for two decades so that he could raise his children. Asked in 2015 what he wanted the first line of his obituary to say, he responded without the slightest hesitation: "I was a great dad."

LUKE'S GIFT

The first two trilogies—Lucas's half dozen—should be called "The Redemption of Anakin Skywalker." The redemption occurs as a result of intense attachment, otherwise known as love. That form of attachment is the whole reason for Anakin's descent to the Dark Side. It's why he falls; he can't bear to lose his beloved. Anakin's heart is what gets him into trouble.

Attachment is also the reason for his choice to return to the Light. He can't bear to see his son die. In the end, Star Wars insists that you can't be redeemed without attachment. That's the strongest message of the saga, and that's what makes it speak to people's deepest selves.

Redemption has everything to do with forgiveness. If you are forgiven, most of all by yourself, you can be redeemed. Luke forgives his father. (A good lesson for children everywhere: if Luke can forgive the worst person in the galaxy, just about, then surely any parent can be forgiven. A good lesson for grudge holders, too. Let it go.) Even toward the end, Luke is willing to give Vader that supreme gift. He redeems him through that forgiveness. As Lucas said, "Only through the love of his children and the compassion of his children, who believe in him, though he's a monster, does he redeem himself."

Star Wars isn't limited to any particular religion. But in that respect, it can claim to be a genuinely Christian tale.

In *The Force Awakens*, Han has precisely the same attitude toward Kylo, his son, as Luke did toward his father, Kylo's grandfather. True, that didn't work out so great. But just wait. For the third trilogy, I predict that some redemption is on the way, and for more than one character. (You'll see.)

"BUT YOU'LL DIE"

In *A New Hope*, Anakin is the satanic figure, the embodiment of evil. He is made good because his son insists on seeing good in him, and chooses to loves him, and because in the end, he chooses to love him back. Martin Luther King Jr. had something relevant to say: "There is some good in the worst of us and some evil in the best of us. When we discover this, we are less prone to hate our enemies." Personal reconciliation often comes from recognizing that truth. The same is true in the political domain; when people who hated each other come together, or when oppressor and oppressed become fellow citizens, that's why. Nelson Mandela understood that well.

The redemption scene is preceded by a vicious fight between father and son. (Every son wants that, a lot, and also hates and is terrified by the idea.) Vader ought to win, as he did in *The Empire Strikes Back;* he is far bigger and appears stronger. But trained by Yoda, Luke succeeds in gaining the upper hand. Vader is forced back, losing his balance, and he is knocked down the stairs. Luke stands at the top, ready to attack. On the verge of victory, he refuses to do so. With his soft, youthful voice he says, "I will not fight you, Father." In his menacing baritone, Vader responds, "You are unwise to lower your defenses." As Vader senses that Luke has a sister, he threatens her with a kind of finality: "Obi-Wan was wise to hide her from me. Now his failure is complete. If you will not turn to the Dark Side, then perhaps she will."

It is at that stage that Luke falls into a Dark Side rage, using his anger, and slashes off his father's right hand at the wrist (a kind of emasculation). Vader is at his son's mercy. The Emperor to Luke: "Good! Your hate has made you powerful. Now, fulfill your destiny and take your father's place at my side!" But rejecting what he himself is becoming, Luke refuses to commit patricide: "You've failed, Your Highness. I am a Jedi, like my father before me." It is then that the Emperor tries to kill Luke, hurling lightning bolts at him. Christlike, Luke asks, "Father, please. Help me." At the last possible moment,

Vader lifts the Emperor and hurls him to his death, saving his son; but Vader himself is dying.

Here's the redemption scene:

DARTH VADER: Luke . . . help me take this mask off.
LUKE: But you'll die.
DARTH VADER: Nothing . . . can stop that now. Just for once . . . let me . . . look on you with my *own* eyes.
[*Luke takes off Darth Vader's mask one piece at a time. Underneath, Luke sees the face of a pale, scarred, bald-headed old man—his father, Anakin. Anakin sadly looks at Luke but then gives a tired smile.*]
ANAKIN: Now . . . go, my son. Leave me.
LUKE: No. You're coming with me. I'll not leave you here, I've got to save you.
ANAKIN: You already . . . have, Luke. You were right. You were right about me. Tell your sister . . . you were right.
[*Anakin smiles and his eyes begin to droop, slumping down in death while giving one last dying breath.*]

For a fairy tale, that's good. Actually, it's very good. It's even great. And a nice bit from the novelization: "The boy was good, and the boy had come from *him*—so there must have been good in *him*, too. He smiled up again at his son, and for the first time, loved him. And for the first time in many long years, loved himself again, as well." (One reason we love other people is that they help us to love ourselves. Luke did that for Anakin, and Han tried to do it for Kylo.)

The sheer quality of the dialogue here is a bit of an upset. Lucas knows myth, and he has a spectacular visual imagination, but most of the time, emotions are not exactly his strong suit. While he enjoys editing, he doesn't always like working with people. (In the prequels, it's droids and more droids. Droid armies everywhere. All droids, all the time.) Harrison Ford famously told him, "You can type this shit, George, but you can't say it." He confesses that he struggles with dialogue. He has said, "I think I'm a terrible writer." Once he admitted, "I'd be the first person to say I can't write dialogue. . . . I don't particularly like dialogue, which is part of the problem." And as Ford remarked in an interview, "George isn't the best at dealing with those human situations—to say the least."

But at the crucial moment in the original trilogy, Lucas delivered. He was the best at dealing with that particular human situation. And he knew exactly what he was doing. On this count, he didn't trick anyone.

Lucas had a lot of sources; for Luke's journey, his major one was Campbell's tale of *The Hero with a Thousand Faces*. The whole series tracks Campbell's account. But the notion of a father sacrificing himself, and repudiating the cause of his entire life, and dying, in order to save his son? That's Lucas's own. It's highly original.

That's what tops "I am your father."

"ATTACHMENT IS FORBIDDEN"—NOT!

The prequels are ostensibly about one thing above all: the perils of attachment. In Anakin's own words: "Attachment is forbidden. Possession is forbidden." Yoda's words: "Let go of fear, and loss cannot harm you." Evidently influenced by Buddhism, Lucas self-consciously portrayed a person turning to evil because he could not "let go"—of his mother and of his beloved. Fear of loss is Anakin's downfall. And of course this theme looms large in Luke's story as well. Luke is reckless and vulnerable to the Dark Side, because he is terrified of losing the people he loves. "His friends were in terrible danger, and of course he must save them."

Once more, Yoda's words: "Train yourself to let go of everything you fear to lose." Yoda again: "Of the Dark Side, despair is." The reason? "Even despair is attachment; it is a grip clenched upon pain."

The point here is plain: If you are attached to someone, you become vulnerable. Recall Yoda's famous words: "Fear leads to anger, anger leads to hate, hate leads to suffering." Serene detachment is the best path, and the only safe one, because it prevents catastrophic choices. Luke nearly fails as a Jedi Knight because of his rage, produced by Vader's vow to pursue his sister. Anakin does fail as a Jedi Knight because he is incapable of detachment; he is desperate to find a way to bring his loved ones back to life. As Lucas put it, "His undoing is that he loveth too much."

The Sith get their revenge only because of Anakin's fear of death— not his own, but of the people he loves. Anakin chooses disastrously as a result of that fear. And in fact, distinguished strands of both Western and Eastern philosophy argue strongly in favor of detachment. Both Stoicism and Buddhism make pleas for detachment, and in emphasizing the perils associated with fear of loss, the Star Wars movies borrow heavily from those traditions. As the philosopher Martha Nussbaum notes, "The Stoics think you should never mourn," and "Cicero reports that a good Stoic father says, if their child dies, 'I was always aware that I had begotten a mortal.'"

But in *Return of the Jedi*, Anakin is redeemed not by serenity and distance but by their opposite. He chooses to kill the Emperor because he cannot bear to see his son die. (So much for those silly Stoics. Chose *that*, Anakin did.)

Whatever Yoda said, Anakin ends up being redeemed by fear of loss and by love, not detachment—and so he is, when making the redemptive choice, perfectly continuous with his earlier self, showing the very characteristics that led him to the Dark Side. When Lucas pressed that point, the Force was unquestionably with him. In terms of narrative, that's his finest moment.

The Redemption of Anakin Skywalker transcends any individual's personal struggles. Its real theme is universal. By their innocence and goodness, by their boundless capacity for forgiveness, and by the sheer power of their faith and hope, children redeem their parents, bringing out their best selves. And as every child knows, deep in his heart, any parent is likely to choose to risk his life to save his child's, even if it means a contest with the Emperor himself. When he makes that choice, the Force is going to be right there with him.

I like that, and I believe it.

FREEDOM OF CHOICE

It's Not About Destiny or Prophecy

*Everybody has the choice of being a hero or not being a
hero every day of their lives. You can either help somebody,
you can be compassionate toward people, you can treat
some people with dignity. Or not.*

—GEORGE LUCAS

In the national anthem of the United States, the most famous words
are these: "Oh, say does that star-spangled banner yet wave / O'er the
land of the free and the home of the brave?" And of those, the most
important are "the land of the free." (Listen to it as if it were fresh
and new.)

A New Hope was made in the United States, in the aftermath of
the 1960s and in the shadow of the civil rights movement, the Soviet
Union, and the Watergate era. Freedom was the national ideal, and it
seemed at grave risk. The civil rights movement suggested that many
Americans were not truly free. When *A New Hope* was released, Martin Luther King's words were hardly a distant memory:

And so let freedom ring from the prodigious hilltops of New Hampshire.

Let freedom ring from the mighty mountains of New York.

Let freedom ring from the heightening Alleghenies of Pennsylvania.

Let freedom ring from the snowcapped Rockies of Colorado.

Let freedom ring from the curvaceous slopes of California.

But not only that:

Let freedom ring from Stone Mountain of Georgia.

Let freedom ring from Lookout Mountain of Tennessee.

Let freedom ring from every hill and molehill of Mississippi.

From every mountainside, let freedom ring.

To many people, including George Lucas, the Nixon administration was a quite serious threat. It was willing (eager!) to punish political adversaries by abusing the tax system, to create an "enemies list," to bribe and to threaten, even to wiretap the Democratic Party. Perhaps the United States would give up on freedom itself? Perhaps it would do so voluntarily? In August 1974, Nixon resigned in the midst of an impeachment effort that would almost certainly have led to his removal from office. The shadow of the Nixon administration loomed over the writing and reception of Star Wars.

Of course, the continuing struggle between the United States and the Soviet Union was a crucial background fact. It was not until 1983 that Ronald Reagan would use the words "evil empire" (as I have noted, he might have been influenced by Star Wars). But in the 1970s, many Americans saw Eastern Europe as a kind of prison, and they regarded the United States as a beacon of freedom, fighting a Cold War for which victory was far from certain. Even if they regarded the Nixon administration with alarm, they believed that American institutions prized freedom and represented a precious and fragile achievement. Maybe communism would prevail?

Opposing the Rebellion to the Empire, Star Wars celebrates political freedom. That is one of the central distinctions between the Light and Dark Sides of the Force. But Star Wars is also after something that is both bigger and far more intimate. It is making a point about the human condition. It is saying something about individual lives, not just political institutions. It is taking an unequivocal stand on a deep philosophical debate. It is urging that freedom of choice is, in a sense, inviolate, whether or not we know it.

Your father might want you to go into business with him, and your Yoda might want you to stay with him in Dagobah, and your captain in the First Order might tell you to report for duty. But really, it's up to you. Sure, the price of exercising your freedom might be high. Even so, it's up to you.

LIBERTY AND AUTOPILOT

Lawrence Kasdan has this to say about Star Wars: *"Force Awakens, New Hope, Empire*—these are movies about fulfilling what is inside you. That's a story that everybody can relate to. Even when you get to be my age, you're still trying to figure that out. It's amazing but it's true. What am I, what am I about, have I fulfilled my potential, and, if not, is there still time? That's what the Star Wars saga is about."

At least as much as the special effects and the creatures and the soaring music, Kasdan's account helps explain the exhilaration that the movies produce. Think about Luke, Han, Anakin, Rey, Finn, and Kylo. At key moments, every one of them chooses to ask Kasdan's four questions. Luke accompanied Obi-Wan to Alderaan, and Han saved Luke's ass, and Anakin killed the Emperor. Finn deserted the First Order, and Rey claimed Luke's lightsaber, and Kylo—well, not so good, but hey, it was still a free choice. (Even the Sith respect freedom when people are choosing between the Light and Dark sides. If you're going to go Sith, you have to do that on your own.)

Whether farm boys or scavengers or Stormtroopers, people live much of their lives on autopilot, as if they are stuck in their situations—a farm, a quarrel, a bad relationship, a bad job. They don't often ask Kasdan's questions. They should. (You should.) The questions are liberating, and just asking them can change everything. Finn certainly found that, and in a different way, so did Han. The movies are a lot of fun, but that's a serious point. Bruce Springsteen, in a "Long Time Comin'": "Yeah I got some kids of my own / Well if I had one wish for you in this god forsaken world, kid / It'd be that your mistakes will be your own / That your sins will be your own."

That we always have freedom of choice is one of the most important things that Lucas meant to say, and his successors are saying it, too. The people who love Star Wars hear it, loud and clear. The point here is sweet and simple, and so we can be brisk.

PHILOSOPHY

There is, of course, an elaborate academic debate over the existence of free will. Are people free to choose, or should we endorse determinism, which insists that they are not?

Some of this debate is empirical; it asks to what extent people's choices are determined by their environment. A great deal of behavioral science suggests that the environment really does matter: whether you will save or spend, lose weight, act fairly, get a job, or even be happy can depend on small features of the social context. You're living amid a kind of choice architecture, and it can have a decisive effect on what you choose. (Weather provides choice architecture, and so do noises and colors, and font sizes, too.) Can we say that same thing about whether you will fall in love with someone, or go to the Dark Side, or kill or instead save your father? That would seem to be an extreme position. But it might be true.

Within philosophy, the debates are more conceptual. On one view, free will exists when we make choices that fit with our own deepest values—those that we endorse after reflection. If you decide to become a doctor, to quit smoking, or to be a bit kinder in the workplace, you are probably exercising your free will. The philosopher Harry Frankfurt makes a distinction between what we want (a cigarette, a bit of extra sleep, a visit to the Dark Side) and what we want to want (no cigarettes, a little more work, the Light). Frankfurt argues that if we act in accordance with what he calls our second-order desires, we are exercising our free will. For Frankfurt, freedom has a lot to do with the ideal of self-mastery.

That's a controversial view, and some people reject it. With respect to the philosophical debates, Star Wars has two things to say. First, Frankfurt is essentially correct. At the key moments, its heroes act in accordance with their deepest values. (That's how they answer Kasdan's questions.) They choose to do so freely; that's how they express their freedom; that's what makes them free. Second, free will is real. At every moment—whether small or large—you get to decide what to do with your own life. You can leave a relationship or a job; you can help someone; you can try to save a life (even your own).

Movies about Luke Skywalker and Obi-Wan Kenobi can't settle long-standing academic debates, but the message is clear. By the way, it's also right.

"ANAKIN SAYS YES AND LUKE SAYS NO"

A lot of people disparage the Star Wars prequels, and understandably so; they're not nearly as good as the original trilogy. But in their own way, they're not just beautiful; they're also awfully clever. Here's the best part: all of the choices in the first trilogy are precisely mirrored in the prequels. The two trilogies are about freedom of choice under nearly identical conditions. Lucas was entirely aware of this: "Luke is faced with the same issues and practically the same scenes that Anakin is faced with. Anakin says yes and Luke says no."

In *Attack of the Clones*, Anakin's visions of his mother's suffering, paralleling Luke's own visions, lead him to choose to disobey his mandate to protect Padmé. He travels home to save his mother, ignoring in the process his stepfather's advice that he accept her death. Unable to rescue her, he makes a literally fatal choice, which is to draw his lightsaber and slaughter those he holds responsible.

In *Revenge of the Sith*, Anakin, like Luke, must decide the fate of a defeated enemy. Luke spares Anakin; Anakin takes a different path. Early on, he decides, at the Emperor's urging, to kill Count Dooku. If you're in the right mood, it's a searing scene. Palpatine: "Good, Anakin, good. I knew you could do it. Kill him. Kill him now!" Anakin responds, weakly: "I shouldn't." But he does, pleading, "I couldn't stop myself." A point for Frankfurt: Anakin is in the grip of his first-order desires, and he can't help it, even though his second-order desire is otherwise: "I shouldn't."

In the pivotal scene in the prequels, the situation in *Return of the Jedi* is explicitly inverted, as Anakin saves Palpatine (Darth Sidious) and ultimately allows him to kill Mace Windu. When Windu is triumphing over Palpatine, the Sith Lord begs for Anakin's help, offering these defining words: "You must choose." He does, choosing Palpatine—and then yields to the Dark Side. The culmination:

ANAKIN SKYWALKER: [after [the death of] Mace Windu and in disarray] What have I done?

DARTH SIDIOUS: You are fulfilling your destiny, Anakin. Become my apprentice. Learn to use the Dark Side of the Force. There's no turning back now.

ANAKIN SKYWALKER: I will do whatever you ask. Just help me save Padmé's life. I can't live without her. If she dies, I don't know what I will do.

DARTH SIDIOUS: To cheat death is a power only one has achieved through centuries of the study of the Force. But if we work together, I know we can discover the secret to eternal life.

ANAKIN SKYWALKER: I pledge myself to your teachings. To the ways of the Sith.

DARTH SIDIOUS: Good. Good! The Force is strong with you, Anakin Skywalker. A powerful Sith you will become. Henceforth, you shall be known as Darth . . . Vader.

ANAKIN SKYWALKER: Thank you . . . my Master.

DARTH SIDIOUS: Lord Vader . . . rise.

At the crucial moments, destiny and prophecies are just background noise. Sidious speaks of destiny, but it's clear that Anakin has made his choice ("I pledge myself to your teachings"). Time and again, the most important characters in Star Wars encounter two paths, and they intuit something about the consequences of both, and they decide accordingly. Padmé insists: "There's always a choice." Does Anakin hear the echo of her voice decades later, when he decides to save their son from the Emperor? I like to think so.

"YOU GET MANY OPPORTUNITIES TO KEEP YOUR EYES OPEN"

Here's Leia, speaking of Han's apparent desertion of the rebellion in *A New Hope:* "A man must follow his own path. No one can choose it for him." Here's Obi-Wan to Luke, again in *A New Hope:* "Then you must do what you think is right, of course." Here are Lucas's own words: "Life sends you down funny paths. And you get many opportunities to keep your eyes open." He was talking about his own life, but he might as well have been talking about Star Wars and the characters who populate it.

In the original trilogy, Darth Vader tells Luke: "It is your destiny, join me and together we can rule the galaxy as father and son."

Wrong! The Emperor tells Luke: "It is unavoidable. It is your destiny. You, like your father, are now . . . mine." Wrong again!

Choices are what doom and redeem Anakin, and they are certainly what turn Han into a fighter for the rebellion (kind of), and Luke into a Jedi. Choices are what turn Finn into a Resistance fighter and Rey into a Jedi-to-be. Lucas once more: "You have control of your destiny. . . . You have many paths to walk down."

In a 2015 interview, here's how Kasdan put it:

> My favorite line that I ever wrote is in *Raiders [of the Lost Ark]*. Sallah says to Indy, "How are you going to get the box back?" And Indy says: "I don't know. I'm making this up as I go." That is the story of everybody's life. It happens to be very dramatic for Indiana Jones. Get on the truck, get on the horse. But for you and me, we're making it up, too. Here's how I'm going to behave. Here's what I'm willing to do to make a living; here's what I'm not willing to do. How we make up our lives as we go. That's such a powerful idea, because it's very exciting.

Yup.

REBELS

Why Empires Fall, Why Resistance Fighters (and Terrorists) Rise

Finn: Han Solo, the rebellion general?
Rey: No! The smuggler!

Star Wars isn't a political tract, but it has a political message. After all, it opposes an Empire to a Republic, and a First Order to a Resistance, and its heroes are rebels, who want to return peace and justice to the galaxy.

That's one reason for the universal appeal of the saga. Whatever your political convictions, and wherever you live, you're likely to see an Emperor of some kind, and you're likely to have some sympathy for the rebels or the Resistance. Your teacher or your boss might seem like an Emperor. Maybe your nation's leader reminds you of Palpatine; maybe the opposing party is the Resistance. You might follow a Skywalker. (In the United States, many people saw John F. Kennedy as a kind of Luke, and Reagan and Obama, too.)

George Lucas certainly had political ideas in mind. I have noted that he modeled Emperor Palpatine on Richard Nixon, and Vietnam provided a relevant backdrop for his tale. In his own words:

> I started to work on *Star Wars* rather than continue on *Apocalypse Now*. I had worked on *Apocalypse Now* for about four years and I had very strong feelings about it. I wanted to do it, but could not get it off the ground. . . . A lot of my interest in *Apocalypse Now* was carried over into *Star Wars*. . . . I figured I couldn't make that film because it was about the Vietnam War, so I would essentially deal with some of the same interesting concepts that I was going to use and convert them into space fantasy, so you'd have essentially a large technological empire going after a small group of freedom fighters or human beings . . . a small independent country like North Vietnam threatened by a neighbor or provincial rebellion, instigated by gangsters aided by empire. . . . The empire is like America ten years from now, after Nixonian gangsters assassinated the Emperor and were elevated to power in a rigged election; created civil disorder by instigating race riots aiding rebel groups and allowing the crime rate to rise to the point where a "total control" police state was welcomed by the people. Then the people were exploited with high taxes, utility and transport costs.

It's pretty safe to say that Star Wars criticizes centralized authority, and its rebel heart lies with those who try to resist it. Later Lucas suggested that he made *A New Hope* "during a period when Nixon was going for a third term—or trying to get the Constitution changed so he could go for a third term—and it got me to thinking about how democracies turn into dictatorships. Not how they're taken over where there's a coup or anything like that, but how the democracy turns itself over to a tyrant." (Actually Nixon was never going for a third term, or trying to change the Constitution, but Lucas is a good storyteller.)

More recently, Lucas described a visit in Europe, after the release of *Revenge of the Sith,* "with a dozen reporters, and the Russian correspondents all thought the film was about Russian politics, and the Americans all thought it was about Bush. And I said, 'Well, it's really based on Rome. And on the French Revolution and Bonaparte.'" The prequels focus on the rise of tyranny and the collapse of democracies. They explore the kinds of machinations that allow dictators to come to power, and they show how republics fall prey to them.

There's a stylized account of the loss of freedom, which Padmé nicely captures: "So this is how liberty dies . . . with thunderous applause." (We'll get to Nazi Germany pretty soon.) With respect to politics and the death of republics, Star Wars tells a perfectly recognizable tale. It offers a warning about the need for citizen vigilance against the countless would-be emperors who try to accumulate power at the expense of the public. That's why people in so many different nations can appreciate its politics—and it will forever be so.

The Force Awakens, J. J. Abrams explained in a magazine interview, "came out of conversations about what would have happened if the Nazis all went to Argentina but then started working together again? What could be born of that? Could the First Order exist as a group that actually admired the Empire? Could the work of the Empire be seen as unfulfilled? And could Vader be a martyr? Could there be a need to see through what didn't get done?" That's the origin of the third trilogy.

DISCUSSING INVASIONS IN COMMITTEE

Star Wars is obsessed with the separation of powers. It speaks of republics and empires, but it's really opposing democratic systems to fascist ones. That's a central theme in the prequels, but you can find it in the first trilogy as well. What are the constraints on the executive branch, and on Chancellors (also known as presidents)? Under what conditions will an executive official claim supreme authority? Isn't the legislature the most democratic branch? Does it fail for that reason? When does it fail?

Emperor Palpatine is able to rise to power only because of the ceaseless, pointless squabbling of legislative representatives in the Republic. He seizes authority as a direct result of that squabbling. (For some Americans in the twenty-first century, witnessing similar squabbling, that seizure of authority does not lack appeal.) Padmé sees the problem: "I was not elected to watch my people suffer and die while you discuss this invasion in committee." So does Anakin: "We need a system where the politicians sit down and discuss the problem, agree what's in the interest of all the people, and then do it." Padmé wonders: "What if they don't?" Anakin: "Then they should be made to."

"A DECREE FOR THE DISSOLUTION
OF PARLIAMENT"

There's a politics to this exchange. The Star Wars saga has a real problem with the concentration of government power in one person. The movies are consistent on this point. An observation at a key moment in the Emperor's rise to power in *Revenge of the Sith:* "The Senate has surrendered so much power; it's hard to see where his authority stops." In *A New Hope*, General Tarkin reports, "The Imperial Senate will no longer be of any concern to us, gentlemen. I have just received word that the Emperor has permanently dissolved that misguided body."

When drafting the prequels, Lucas began researching the transition from democracies to dictatorships, examining "why . . . the senate after killing Caesar turn[ed] around and g[a]ve the government to his nephew? . . . Why did France after they got rid of the king and that whole system turn around and give it to Napoleon?" He noted:

> It's the same thing with Germany and Hitler. . . . You sort of see these recurring themes where a democracy turns itself into a dictatorship, and it always seems to happen kind of in the same way, with the same kinds of issues, and threats from the outside, needing more control. A democratic body, a senate, not being able to function properly because everybody's squabbling, there's corruption.

Hitler was apparently a model for Palpatine, and in Germany, the fuhrer's own rise was confirmed by his successful claim to general authority to make law, free from any requirement of legislative authorization. In the midst of an apparent crisis, signaled by a fire at the Reichstag (legislative) building, Hitler demanded that authority. A chilling newspaper account on February 2, 1933, reads like something right out of Star Wars, but it's quite real:

> The power to dissolve Parliament at his discretion and to rule Germany by decree without Parliament was entrusted

today to Adolf Hitler, Germany's new Chancellor by President Paul von Hindenburg, according to the *Deutsche Allgemeine Zeitung*, an organ close to the Government. President von Hindenburg signed a decree for the dissolution of Parliament which is expected to become effective before the reconvening of Parliament, scheduled for next Tuesday.

In *Attack of the Clones*, Mas Amedda says: "The Senate must vote to give the Chancellor emergency powers." Accepting those powers, Palpatine insists, "It is with great reluctance that I have agreed to this calling. I love democracy. I love the Republic. Once this crisis has abated, I will lay down the powers you have given me!" Yeah right.

DELEGATING POWER

Many legal systems, including those of the United States and Germany, impose real barriers to the granting of the power to rule by decree. Under American law, there is something called a "nondelegation doctrine," which is generally understood to forbid Congress from giving the president the authority to do whatever he wishes. Congress cannot authorize the president to rule by decree. It is not permitted to enact a law that says, "The president is hereby authorized to enact such laws as he wishes."

But on occasion, citizens object that the president is doing something like that. Under President George W. Bush, for example, many people contended that waging the war on terror had led the executive branch to assume Empire-like authorities, intruding on personal privacy to promote national security. In their view, President Bush was essentially ruling by decree. In fact Vice President Dick Cheney embraced the Dark Side, kind of, or maybe more than that:

> We also have to work, though, sort of the dark side, if you will. We've got to spend time in the shadows in the intelligence world. A lot of what needs to be done here will have to be done quietly, without any discussion, using sources and methods that are available to our intelligence agencies, if we're going to be successful. That's the world these folks operate in, and so it's going to be vital for us to use any means at our disposal, basically, to achieve our objective.

Notice the key words: "any means at our disposal, basically, to achieve our objective." And in fact, some of the defenders of the Bush administration came close to arguing that when the nation faces a serious security threat, the President is allowed to do whatever he deems necessary to protect against it. Is that the power to rule by decree? It's not so far from that.

Under President Barack Obama, some critics objected that because of a stalled Congress, the executive branch chose to operate as a kind of Palpatine, similarly blaming legislative squabbling as an excuse for imperial authority. Climate change, immigration reform, gun control, economic policy—in these and other areas, President Obama acted when Congress wouldn't. In his own words: "I want to work with Congress to create jobs and opportunity for more Americans. But where Congress won't act, I will." He did.

Is that an admirable insistence on using executive power to help people? Or is it an assertion of Imperial authority? I worked in the

Obama Administration for nearly four years, and I firmly believe the former, but some people definitely disagree.

In the early decades of the twenty-first century, there has been a lot of squabbling in Congress. The influential Senator Dick Durbin, a Democrat, embraced executive unilateralism: "I think there reaches a point where Congress is just being stubborn. They are just opposing everything [the president] suggests, and he has to make decisions in the best interests of the country. . . ." Was Senator Durbin capitulating to an Emperor?

I don't think so, but the Star Wars movies can't possibly answer that question. True, they offer some enduring truths: freedom is good, oppression is bad, and public officials shouldn't torture or choke people. But let's hope that you didn't need Star Wars to know that.

Star Wars does touch on another and subtler point, which involves the nature and the fate of rebellions. Many rebels begin with high ideals, but once they are in power, their idealism fades, and something else takes over. Pragmatism? The quest for power itself? The desire to hold onto it? The French Revolution, which became very bloody, is a prominent case in point. Some of the heroes of the Arab Spring have not turned out to be democracy's friends. Thus Padmé: "What if the democracy we thought we were serving no longer exists, and the Republic has become the very evil we have been fighting to destroy?"

So let's talk about rebellions.

CONSERVATIVE REBELS

What do Martin Luther King Jr. and Luke Skywalker have in common?

They're both rebels, and they're rebels of the same kind: conservative ones. If you say you want a revolution, you might choose to follow them, at least in that regard. Conservative rebels can be especially effective, because they pull on people's heartstrings. They connect people to their past, and to what they hold most dear.

Some people, like Leia Organa, seem to be rebels by nature, and whenever a nation is run by Sith or otherwise evil or corrupt, they might think that rebellion is a great idea. They might well be willing to put their own futures on the line for the cause. But in general, even rebels do not like to "reboot"—at least not entirely. This is true whether we are speaking of our lives or our societies.

Of course some people want to blow everything up and start over. That might be their temperament, and it might be what their own moral commitments require. But human beings usually prefer to continue existing narratives—and to suggest that what is being written is not a new tale but a fresh chapter, a reform to be sure, but also somehow continuous with what has come before, or with what is best in it, and perhaps presaged or foreordained by it. That's true for authors of Episodes of all kinds, and not just Lucases and Skywalkers.

Consider the words of Edmund Burke, the great conservative thinker (and admittedly no rebel), who feared the effects of "floating fancies or fashions," as a result of which "the whole chain and continuity of the commonwealth would be broken." To Burke, that's a tragedy, a betrayal of one of the deepest human needs and a rejection of an indispensable source of social stability. Burke spoke with strong emotion about what would happen, should that break occur: "No one generation could link with the other. Men would become little better than the flies of a summer."

Pause over those sentences. Burke insists that traditions provide connective tissue over time. That tissue helps to give meaning to our lives, and it creates the closest thing to permanence that human beings can get. This is a conservative thought, of course, but even those who do not identify as conservative like and even need chains and continuities. That's part of the appeal of baseball; it connects parents with their children, and one generation to another. The same thing can be said about Star Wars, and it's part of what makes the series enduring. It's a ritual.

In the Star Wars series, what the rebels seek is a *restoration of the Republic*. In that sense, they are the real conservatives. They can be counted as Burkeans—rebellious ones, but still. They're speaking on behalf of their own traditions. By contrast, Emperor Palpatine is the real revolutionary, and so are the followers of the First Order. Luke, the Rebel Alliance, the Resistance want to return to (an idealized version of) what came before. They look backward for inspiration. In fact that's kind of primal.

Martin Luther King Jr. was a rebel, unquestionably a Skywalker, with a little Han and more than a little Obi-Wan. He sought fundamental change, but he well knew the power of the intergenerational link. He made claims of continuity with traditions, even as he helped to produce radically new chapters.

From King's speech about the Montgomery Bus Boycott:

> If we are wrong, the Supreme Court of this nation is wrong. If we are wrong, the Constitution of the United States is wrong. If we are wrong, God Almighty is wrong. If we are wrong, Jesus of Nazareth was merely a utopian dreamer that never came down to earth. If we are wrong, justice is a lie. Love has no meaning.

UNANTICIPATED REVOLUTIONS,
LARGE AND SMALL

In the terrific novelization of *A New Hope*, Biggs, a friend of Luke, is a fairly major character. (He appears briefly in the movie.) Early in the novel, Biggs comes out to Luke, as a rebel wannabe. It's true that he doesn't know what to do, exactly, to rebel. He doesn't even know where the rebel bases are, or if they are, or how to contact them. Here's the key passage:

> "I know it's a long shot," Biggs admitted reluctantly. "If I don't contact them, then"—a peculiar light came into Biggs's eyes, a conglomeration of newfound maturity and . . . something else—"I'll do what I can, on my own."

Biggs has a rebel's heart. And in the world of Star Wars, he's hardly the only one. General Tagge, who has a "certain twisted genius," knows the Empire's challenge: "Some of you still don't realize how well-equipped and organized the Rebel Alliance is. Their vessels are excellent, their pilots better. And they are propelled by something more powerful than mere engines; this perverse, reactionary fanaticism of theirs. They're more dangerous than most of you realize." The key word is *fanaticism*, which can propel ordinary people to do extraordinary things.

Obi-Wan captured the revolutionary sensibility: "Remember, Luke, the suffering of one man is the suffering of all. Distances are irrelevant to injustice. If not stopped soon enough, evil eventually reaches out to engulf all men, whether they have opposed it or ignored it."

Reliable rebels live by that creed. They agree that distance is indeed irrelevant to injustice—and so they choose to combat it.

The idea that evil eventually engulfs us all was captured by the Protestant pastor Martin Niemöller, a prominent critic of Adolf Hitler, who spent seven years in concentration camps:

> First they came for the Socialists, and I did not speak out—Because I was not a Socialist.
>
> Then they came for the Trade Unionists, and I did not speak out—Because I was not a Trade Unionist.
>
> Then they came for the Jews, and I did not speak out—Because I was not a Jew.
>
> Then they came for me—and there was no one left to speak for me.

Political leaders are often surprised and even stunned by rebellions. A long time ago, in a galaxy really far away, Emperor Palpatine had no clue that Luke would resist his entreaties, that Darth Vader would turn on him, and that the rebels would refuse to be broken. In 1770, the British did not foresee the energy and intensity with which the Americans would press their revolution. In 1990, very few people anticipated that in January 1992 the Soviet Union would cease to exist. In 2009, the world could hardly anticipate that the Arab Spring would come just a year later.

The last example is especially revealing, because it is the most recent, and because it caught essentially everyone by surprise.

Notwithstanding the extraordinary intelligence-gathering abilities of many modern governments, no one had any clue about what was coming. For example, the British Foreign and Commonwealth Office admitted its failure to "predict that a spark in Tunisia in December 2010 would trigger such an outpouring of protest." It added: "No other international player, academic analyst, or opposition group within the region foresaw this either." The United States and Canada have acknowledged that their intelligence units likewise missed the movement, and the "vast majority of academic specialists on the Arab world were as surprised as everyone else by the upheavals."

How come? New York University's Jeff Goodwin contends that surprise was essentially inevitable. In his words:

> We know that something like a "revolutionary bandwagon" occurred in Tunisia beginning in December [2010], following a seemingly insignificant event, namely, the self-immolation of a provincial fruit vendor after his business was shut down by the local police. The example of the Tunisian uprising, which culminated in the dictator Ben Ali's precipitant flight from the country, then helped to ignite a revolutionary bandwagon in Egypt, which then soon spread to Libya and other countries where opposition to regimes was widespread and revolutionary thresholds relatively low—although the latter factor in particular could not be known beforehand. The fact that revolution did not spread to Algeria, Saudi Arabia, Jordan, or any number of other Arab countries indicates that the distribution of revolutionary thresholds in those countries was simply not conducive to mass uprisings—although,

again, no one could have foreseen just where and how far the Arab Spring would travel.

That's a bit complicated. Let's try to unpack it.

BLINDNESS

One reason that the Emperor Palpatines of the world are so blind is that they are often insulated and surrounded by terrified lieutenants who offer "happy talk"—assurances that all is well, that everyone loves (or fears) them, and that things are indeed unfolding according to plan. Another reason is that like most human beings, emperors tend to be overconfident and unrealistically optimistic, and their beliefs are affected by their motivations.

In general, human beings tend to believe what they want to believe, and not to believe what they don't want to believe. ("I don't like that and I don't believe that.") Emperors want to believe that people are content rather than angry, or that any anger is limited to a few, or that a rebellion based on anger, if widespread, can be stopped at the point of a gun. If citizens are unhappy, that's an inconvenient truth, which leaders might well ignore. You don't need to be a Sith to believe that any rebellion is destined to fail.

The more puzzling fact is that it is not merely political leaders who fail to anticipate a successful rebellion; in many cases, *almost everyone makes the same mistake.* How can that be?

We have already had a glimpse of one explanation: social dynamics are responsible for how things are received, and it's hard or even impossible to foresee the nature of social dynamics. A cause or an idea can be just like a song, a book, or a movie. People might follow it, and even lay down their lives for it, because of what they think other people think. An attempted rebellion might turn out to be like Sixto Rodriguez in the United States, or like Sugar Man in South Africa. Everything depends on what each of us thinks that the rest of us think.

Recall this exchange in *A New Hope:*

OBI-WAN KENOBI: [*to Luke*] You must learn the ways of the Force, if you're to come with me to Alderaan.

LUKE SKYWALKER: Alderaan? I'm not going to Alderaan, I've gotta get *home*; it's late, I'm in for it as it is!

OBI-WAN KENOBI: I need your help, Luke. She needs your help. I'm getting too old for this sort of thing.

LUKE SKYWALKER: Look, I can't get involved. I've got work to do. It's not that I like the Empire; I hate it, but there's nothing I can do about it right now. . . . It's all such a long way from here.

OBI-WAN KENOBI: That's your uncle talking.

Like many potential rebels, Luke emphasizes three points: 1) He's got work to do. 2) There's nothing that he can do about the Empire. 3) Any relevant action is really far away. Note also that Luke's resistance is pretty weak. He doesn't much like his work, and he actually has some enthusiasm for going to a place that is a long way from home. A good bit from the novel version of *A New Hope:* "Biggs is right. I'll never get out of here. He's planning rebellion against the Empire, and I'm trapped on a blight of a farm."

One question is whether there is, in fact, anything that Luke can do about the Empire. A feeling of hopelessness can certainly dampen engagement. But if the Lukes of the world are assured that other people are engaged in a rebellion, their resistance might dissipate. A great deal might turn out to depend on whether potential rebels see the equivalent of a lot of early downloads.

CASCADES OF REBELLION

Political scientist Susanne Lohmann, of the University of California at Los Angeles, sees serious protest activities as informational cascades with just three steps:

1. People take costly political action to express their dissatisfaction with the incumbent regime.

2. The public takes informational cues from changes in the size of the protest movement over time.

3. The regime loses public support and collapses if the protest activities reveal it to be malign.

That's not the worst imaginable description of how and why the Empire fell in the original trilogy. (Of course it's not the best imaginable description, either, but bear with me for a bit.)

As Lohmann elaborates her model, societies can be sorted into different groups, with different thresholds for taking action. Some people will just rebel, no matter what. They hate the status quo, and they're brave, and they're perfectly willing to change things, even if no one else is. Let's call them the Princess Leias. Luke's friend Biggs is also one of these types: "Luke, I'm not going to wait for the Empire to conscript me into its service. In spite of what you hear over the official information channels, the rebellion is growing, spreading. And I want to be on the right side—the side I believe in."

Others dislike the status quo, but they will rebel only if they reach a certain level of disaffection and rage. Let's call them the Lukes. ("I want to come with you to Alderaan.") Others dislike the status quo, but they will rebel only if they have a sense that the rebellion will actually succeed. They might pretend not to care, or their interests might be purely commercial, but in their heart of hearts, they sympathize with the rebellion. Let's call them the Hans. Still others are apathetic, and their decision whether to rebel, or instead to support the regime, depends on which way the winds are blowing. Let's call them the Naboo. A final group supports the regime, and will do so even as the protests mount. Let's call them the Sith.

On this account, the success of a rebellion very much depends on social dynamics, and on the strength of the signal given by rebels. If the Princess Leias seem numerous enough, and if the Empire seems bad enough, the Lukes and Hans will join them, and if there are enough Lukes and Hans, the Naboo will flip in their direction. Of course the Sith will stick with the regime—they might even *be* the regime—but once they are isolated, they are bound to lose power.

WHAT DO PEOPLE ACTUALLY THINK?

In a truly repressive society—one against which rebellion is most justified—it will be very hard to know the magnitude of people's dissatisfaction, because people *will not say what they really think*. Like Luke, they might hate their leaders and hope for a change, but they will be well aware that if they speak out, they will be at risk. They will falsify both their preferences and their beliefs. No one will know what public opinion actually is; there is a silent majority.

That's one reason, by the way, that opinion polls cannot be trusted in authoritarian nations. People might well say that they are satisfied with their government even when they are profoundly unhappy with it.

A little story: In the late 1980s, I was asked to teach a short course on American law in Beijing. (We didn't discuss Star Wars, so far as I can remember. Recall that Star Wars was not shown in China until 2015.) As a final assignment, I asked my thirty students to write a short paper. Their task was to explore what the United States could learn from the Chinese legal system, or what China could learn from the U.S. legal system. They were free to pick one or the other. I much looked forward to seeing what they would come up with.

To my utter amazement, almost everyone in the class refused to do the assignment! With embarrassment, one of them explained: "We are worried that what we write could get into the wrong hands." By that, they meant to suggest that they could get in trouble with their own government. Of course they were loyal to their country. And in private, they were willing to raise some questions about what their government was doing (as well as about what the United States was doing)—but for fear of some kind of punishment, they were unwilling to put those questions in writing.

Here's the upshot, elaborated at length by the economist Timur Kuran in his terrific 1997 book, *Private Truths, Public Lies*: If people falsify their preferences and beliefs, rebellions will be difficult or perhaps impossible to predict. People might be satisfied with their government; they might dislike it, at least a little; or they might hate it. Because what people say does not match what they think, citizens will be in a situation of *pluralistic ignorance*: They will have no idea what their fellow citizens believe. But if some people (the Leias among them) start to express dissatisfaction and display a willingness to rebel, then others (the Lukes) might think that a rebellion could succeed, because a lot of people might be prepared to join it. If so, the world might turn upside down.

The unpredictability of rebellions has a lot to do with social dynamics associated with cascade effects—but it also comes from not knowing what people really think about the status quo.

GROUP POLARIZATION

We haven't sufficiently discussed the *internal* dynamics of rebellion. What makes people get agitated enough to rebel?

One possibility is that they are truly unhappy or miserable, perhaps because of what their leaders do, perhaps because of what their leaders fail to do. Like Leia and Biggs, they have an acute sense of grievance or injustice. (That was certainly true in the American Revolution, and the French Revolution, too, and the attack on apartheid in South Africa, and the Arab Spring.) Maybe some Empire is responsible for the deaths of their aunts and uncles.

There is no question that general unhappiness (economic distress or a sense of humiliation, exploited by rebel leaders) can provoke a rebellion. It is equally clear that people can be radicalized by particular precipitating events—especially when tyranny strikes close to home. I have mentioned the "availability heuristic," which means that people assess probabilities by asking whether a relevant event comes easily to mind. If a crime occurred recently in your neighborhood, or if someone in your family got cancer, you might have an exaggerated sense of the risk of crime or cancer. Often rebellions are spurred because a particular event becomes easily available to many minds: a killing of an innocent civilian, a jail sentence for a dissenter, an abuse of power by the tax authorities. And when rebellions take off, a big reason is the phenomenon of *group polarization*—which helps to explain not only the rise of republics, and the return of the Jedi, but also the creation of empires in the first place, and the revenge of the Sith.

Group polarization occurs when like-minded people, talking mostly with one another, end up thinking a more extreme version of what they thought before they started to talk. Suppose that Facebook friends are discussing whether President Obama is great or terrible, or whether climate change is a serious problem, or whether J. J. Abrams knocked it out of the park or ruined everything. If most of them start with the thought that Obama is great, that climate change is a serious problem, and that Abrams ruined everything, then the consequence of their conversations will be to make them more unified and more confident, and to hold their original views with more intensity.

A lot of social science research, in many of the world's nations, has shown that this is a consistent pattern. If you put a bunch of rebels in a room and ask them to discuss the rebellion, they'll get more extreme. The American Revolution was spurred in this way, and so too the Reagan Revolution, and Obama's election in 2008. And if any group—whether Jedi or Sith—is asking the important question, *Why do they hate us?* the answer probably has everything to do with group polarization.

It follows that, for example, a group of people who tend to approve of an ongoing war effort will, as a result of discussion, become still more enthusiastic about that effort; that people who were disappointed by *The Force Awakens* will end up more disappointed still if they keep talking about it; that people who think that gun control is greatly needed, and that the United States should have much more of it, will become even more committed if they talk mostly with one another; that people who disapprove of the United States, and are suspicious of its intentions, will increase their disapproval and suspicion if they exchange points of view. Indeed, there is specific evidence of the latter phenomenon among citizens of France. If people in Paris get together, and if they aren't at all happy with the United States, their discussions with each other might well make them intensely anti-American.

By the way, the rise of terrorism has a lot to do with group polarization. Terrorists aren't, by and large, poor, or poorly educated, or mentally ill. It's tempting but mistaken to say that if we eliminated poverty and promoted literacy, we would eliminate terrorism at the same time. Many terrorists aren't poor, and they have plenty of education. (Contrary to that perverse reading of the original trilogy, Luke was hardly a terrorist, but Luke types—young men, smart, aggressive, hanging out with other young men—do fall prey to that particular Dark Side.) Terrorism arises as a result of social networks and in particular echo chambers, in which people talk and listen mostly to one another. Conspiracy theories also tend to arise in this way. But Light Side rebellions are spurred by group polarization as well.

WHY GROUPS POLARIZE

What explains people's movements toward extremism? There are two major answers.

The first is based on the *exchange of information*. People respond to the information held by others and the arguments they make, and the "information pool," in any group with some initial disposition in one direction, will inevitably be skewed toward that disposition. A group whose members tend to think that the Empire is tyrannical, or that the United States is engaged in a general campaign against Islam and seeks to kill and humiliate Muslims, will hear many arguments to that effect. It will hear few opposing arguments, simply as a result of the initial distribution of positions within the group.

If people are listening, they will have a stronger conviction, in the same direction from which they began, simply as a result of discussion. The phenomenon is general. A group whose members tend to think that the Star Wars prequels were lousy will hear a large number of arguments against the prequels (ugh, Jar Jar Binks!) and a fewer number of arguments in their favor (just look at how those ships swoop and swerve!). There is considerable empirical support for the view that the information pool has this kind of effect on individual views. (In fact I think people underrate the prequels for exactly that reason.)

The second explanation has to do with *social influences*. The central idea here is that most people care about what others think of them, and once they hear what others believe, they have a tendency to shift their positions accordingly. Suppose that you find yourself in a group of people who think that Star Trek is much better than Star Wars, or that the United States faces acute and imminent terrorist threats, or that climate change is not nearly as big a problem as some people seem to think. As you learn about other people's views, you might shift your position, at least a bit. In that particular group, you might not want to seem to be stupid or immoral.

It should be clear how social influences can help foment a rebellion. If rebels are speaking mostly to one another, group members won't want to seem like compromisers or dupes of an Empire. Civil rights movements emerge, and draw energy, in just this way. In the 1970s, the feminist movement was spurred by social influences. In the early part of the twenty-first century, unanticipated success by efforts to promote LGBT rights had everything to do with group polarization. When the Supreme Court required states to recognize same-sex marriages in 2015, it was effectively ratifying an emerging social consensus, which group polarization had made possible.

There is a final point. Many people, much of the time, lack full confidence in their views. As a result, they offer a moderate version of what they are inclined to think, for fear of seeming reckless or stupid, or being marginalized or ostracized. Many other people have more confidence than they are willing to show, for fear of being proved foolish; such people moderate their views in public. In either case, group dynamics can push people toward a more extreme position. Once people find their views corroborated by others, they become more confident and thus less moderate. That's one way that rebels are created. In his early days, Luke Skywalker was a case in point.

A DOWN LOOK

Reputational cascades play a large role in rebellions. Some people join them because their friends and neighbors want them to—not because they actually care about them. (Han pretended to be an example.) Some people refrain from rebellion because they don't want to risk their reputations or their lives; they engage in rebellion for exactly the same reason. And of course, network effects may play a big role in getting people to fight empires of all kinds. Once the number of rebels increases, participation can become far more energizing; it can seem like the best club in the history of the world.

Importantly, a lot of people's political preferences are weakly held. They aren't sure what they believe. They might think that the current regime is all right, or even pretty good, but if they get some facts, or a few stories, they can be persuaded that it is really bad. Their preferences might even be a product of the fact that the current political system seems inevitable; there's nothing to be done about it.

For many people, it isn't a lot of fun to think that one's leaders are corrupt or tyrannical, or even merely unfair or incompetent. Even if there isn't a Darth Vader around to choke you, you might think that life is just easier if you live your life as if things are okay, or at least okay enough. For that reason, people are often motivated to think the status quo is acceptable or better.

Describing the hierarchical nature of pre-revolutionary America, the great historian Gordon Wood writes that those "in lowly stations . . . developed what was called a 'down look.'" They "knew their place and willingly walked while gentlefolk rode; and as yet they seldom expressed any burning desire to change places with their betters." In Wood's account, it is impossible to "comprehend the distinctiveness of that premodern world until we appreciate *the extent to which many ordinary people still accepted their own lowliness.*"

The simple point here is that as a rebellion gains steam, people become less likely to accept their own lowliness. That "down look" ceases to be a part of life. Instead it becomes a symbol of oppression. A word to emperors of all kinds: beware.

BUTTERFLIES EVERYWHERE

The great science fiction writer Ray Bradbury produced a famous short story about what has come to be known as "the butterfly effect": if a butterfly had been killed at a certain point in time, might things have unfolded in fundamentally different ways? Might the dinosaurs have lived? To see the intuition, assume that Hitler's mother and father, or Ronald Reagan's, or Barack Obama's, had not been together, or had been particularly tired, on what turned out to be the night of conception. With a little twist, surely one or another parent might well have been diverted. No Hitler, no Reagan, no Obama. And if we put conception to one side, we can easily find a large number of serendipitous occurrences, almost-didn't-happens, that were necessary for each of the three.

Because causal chains are so complex, and because so many events are necessary conditions for others, the idea of a butterfly effect is not at all preposterous. If someone's dog had gotten sick on an important occasion, or if someone else had stayed home rather than going out, or sent an email at a particular time, perhaps everything would have been different. World-changing butterflies are everywhere; they define our lives.

Consider the case of George Lucas, who badly wanted to go to art school, but who was strongly discouraged from doing so by his father. He had been planning to go to San Francisco State to major in anthropology. But John Plummer, a childhood friend who was attending the University of Southern California, suggested that he should apply to the new film school there, emphasizing, "they've got a photography school . . . you'll love it. It's easier than PE." He convinced Lucas to take the admissions test, and he got in. For Lucas, that changed everything. As he acknowledges, "I got there on a fluke." And when he got there, he was surprised to find that "it was a cinema school. And I said: 'What! You mean you can go to college to learn to make movies? This is insane!'"

Without John Plummer and that particular conversation: no Star Wars.

In an influential essay in 1972, Edward Lorenz, a meteorologist, made a systematic argument on behalf of the butterfly effect. His essay was called "Predictability: Does the Flap of a Butterfly's Wing Set Off a Tornado in Texas?" Lorenz's thesis was grounded in his observation that with a seemingly trivial variation in a computer simulation of weather patterns, the long-term forecast could be massively changed. In principle, a flap of a butterfly's wings in Mexico could produce a big shift in weather patterns in Texas.

The larger lesson is that because natural and social orders are interacting systems, and because the effects of seemingly tiny alterations can be huge, accurate predictions might be difficult or even impossible. In the words of forecasting expert Philip Tetlock, a woman in Kansas might be surprised to find that "one obscure Tunisian's actions led to protests, that led to riots, that led to a civil war, that led to the 2012 NATO intervention, that led to her husband dodging antiaircraft fire over Tripoli."

Something a bit like this certainly defines the arc of Star Wars. In *A New Hope*, Han Solo chooses to leave the vulnerable rebellion and to go off on his own. He's a Solo, after all, not part of a team. Princess Leia hates that. She wishes it were otherwise, but she notes, "He's got to follow his own path. No one can choose it for him." (Again, the constant Star Wars theme.) Of course he elects to return at the crucial moment, saving Luke from his father, who is about to kill him. So he's not so much of a Solo after all. If Han hadn't made that particular choice, Luke and the rebellion would have ended badly.

Speaking of Han Solo, Harrison Ford was mostly a carpenter, not an actor, until the ripe age of thirty-five. When Lucas was auditioning people for Star Wars, he happened to see Ford doing some woodworking at the studio. Ford was there because Fred Roos, a casting director, hired him to work on a new door. Lucas knew Ford; he had a small part in Lucas's previous film, *American Graffiti*. But Lucas had vowed not to use any of the actors from that film in his "little space thing." Nonetheless, he just happened to see the carpenter-actor in the studio at the time and decided to give him a try as Han Solo.

It's impossible to imagine *Star Wars* without Ford as Solo, but it almost happened. In Roos's words: "Harrison had done a lot of carpentry for me; he needed money, he had kids, he wasn't a big movie star yet. The day he was doing it, George happened to be there. It was serendipitous."

A VERY QUICK GLANCE AT POLITICAL CAMPAIGNS

For many people, *Attack of the Clones* is the least successful, and the very worst, of the Star Wars movies. (But I like it! Have another look? At least check out the amazing opening scene?) Nonetheless, its crawl says something pretty smart about politics and political campaigns:

> There is unrest in the Galactic Senate. Several thousand solar systems have declared their intentions to leave the Republic. This separatist movement, under the leadership of the mysterious Count Dooku, has made it difficult for the limited number of Jedi Knights to maintain peace and order in the galaxy. Senator Amidala, the former Queen of Naboo, is returning to the Galactic Senate to vote on the critical issue of creating an ARMY OF THE REPUBLIC to assist the overwhelmed Jedi. . . .

Here's a way to understand what the crawl is describing. The galaxy is in the midst of a cascade, in which solar systems are not acting independently, but are following one another. Once a few leave the Republic, others follow, and once others follow, the cascade gets bigger, and the pressure for secession increases. Count Dooku is well aware of this fact, and he is trying to exploit it. The Jedi are overwhelmed by rising unrest, which breeds further unrest. Senator Amidala is hoping not only to create an ARMY OF THE REPUBLIC but also to do something about the growth of the cascade.

In the electoral domain, cascade effects are crucial, making or breaking contenders in short periods. In 2008, Barack Obama was clearly the beneficiary of both informational and reputational cascades. As in the case of *A New Hope,* his popularity begat more popularity, and his proven ability to raise money made it easier to attract other donors. In 2015, Republican Scott Walker (remember him?) was the victim of a negative cascade. Early on, Walker was thought to be a front-runner for the Republican nomination, and a lot of smart people thought that he would be the next president. But once people started to say that he was a loser, the idea that he was a loser started to spread, and things snowballed from there. Job applicants find themselves with no offers just because they've been rejected before; so too, once people saw that others were not giving Walker much money, it became even harder for him to raise funds. His campaign collapsed for that reason. (Countless promising candidates have had similar experiences, and countless will in the future.)

Of course, Obama's successes and Walker's failures are not solely the products of cascade effects. Obama was an exceedingly strong campaigner. Walker was surprisingly weak. But it is impossible to understand the stunning successes of Obama, and Walker's collapse, without reference to the same kinds of cascade effects from which successful rebellions, and Star Wars itself, have benefited. For victorious politicians, group polarization is important, too. A real trick is to get one's supporters in a lot of rooms together, even if those rooms are only online. People stir each other up to higher levels of excitement, giving their money and their time.

By the way, an appreciation of those effects also helps to show why national polls are much less important than many experts think. Because the early primaries create cascades, it does not greatly matter if a particular candidate is ahead in those polls by five points, or ten points, or fifteen points—except insofar as such differences help to create early cascade effects. In the United States, candidates would do well to have this thought: I am running to be president of Iowa (and then New Hampshire).

When campaigns begin, as when movies and books are released, it is especially interesting to see how cascade effects, whether negative or positive, start to accelerate. By their very nature, such accelerations cannot be predicted in advance, but once they are under way, they are unmistakable. At a certain stage, what seemed wildly improbable or at best speculative becomes all but certain, as both donors and voters settle on one or another candidate in massive numbers.

After they flock, many observers will insist that the outcome was an inevitable product of the successful candidates' biography, virtues, and ideas, or that their success had some deep connection with the culture or the zeitgeist. But as with blockbuster movies, the impression of inevitability will be an illusion. In all probability, the eventual nominees will have won mostly because of their ability to do what George Lucas and Princess Leia did, which is to manage, and to spark, favorable cascades.

THE OBJECTIVE, AUTHORITATIVE RANKING
OF STAR WARS MOVIES

Enough about politics. Do social influences help account for evaluations of Star Wars movies?

There is no question that with those movies, as with so much art and entertainment, both critics and ordinary people tend to polarize in one or another direction. We see both "up" cascades and "down" cascades. When *The Force Awakens* opened, there was an immediate sense that Abrams had done something amazing and spectacular. That sense was fortified by the fact that early enthusiasts were speaking mostly with each other. Maybe Abrams had made the best Star Wars movie ever! Or at least the second best, after *The Empire Strikes Back*?

After a time, the predictable backlash occurred, as some smart people, and some killjoys, pointed out that *The Force Awakens* borrows a great deal from *A New Hope* and *Return of the Jedi*, and that it seemed to lack originality and boldness, and also Lucas's distinctive genius and willingness to take risks. (Lucas himself added some fuel to the fire, pointing to its lack of originality.) So there was a process of group polarization against *The Force Awakens*—and in some circles, in the direction of reassessing the prequels, and evaluating them far more favorably.

My own view is that with respect to *The Force Awakens*, Abrams did great. It moves fast, and it's a ton of fun. Rey in particular is fabulous. There's nothing bad or embarrassing about it, and it offers some terrific new mysteries. Improbably, it succeeds in blending old and new. That took immense skill. True, it's pretty much a remake. That's okay; a relaunch can be a remake. It doesn't have anything like Lucas's originality, but it's still awfully good.

Let's end this discussion of rebellions and social influences with an objective, authoritative ranking of the movies, wholly unaffected by any such influences:

1. *The Empire Strikes Back* (grade: A+)

2. *A New Hope* (grade: A+)

3. *Return of the Jedi* (grade: A)

4. *Revenge of the Sith* (grade: A-)

5. *The Force Awakens* (grade: A-)

6. *Attack of the Clones* (grade: B-)

7. *The Phantom Menace* (grade: C+)

Sure, it's possible to raise questions. *The Empire Strikes Back* and *A New Hope* unquestionably belong at the top, but *A New Hope* was the first, and the most original, and it set everything in motion. It also poses all those mysteries. Maybe it belongs on top? Fair to ask, but *The Empire Strikes Back* deepens everything, and "I am your father" can be found there. Almost every scene crackles. Besides, it has Han respond "I know" after Leia says "I love you," and it features those amazing four-legged walkers, "the most heavily armored ground vehicles in the Imperial Army." (Technical name: The All Terrain Armored Transport.)

As among *Return of the Jedi*, *The Force Awakens*, and *Revenge of the Sith*, it's a close-ish call. *The Force Awakens* is the tightest and the least flawed of the group, so it wouldn't be entirely nuts to rank it as high as third—wrong, but not entirely nuts. *Revenge of the Sith* has some amazing moments. As we've seen, the inversion of *Return of the*

Jedi is ingenious, and it might be the most impressive of all in terms of the visuals. Anakin's turn to the Dark Side is searing; the final battle between Obi-Wan and Anakin is terrific. As between *The Force Awakens* and *Revenge of the Sith*, we've got a near-tie, and the latter wins by just a whisker. *Return of the Jedi* could have been shorter (it has too much filler), and so people could be forgiven for asking whether *The Force Awakens* and *Revenge of the Sith* might be better. But in the end, the answer is clear. At its best, *Return of the Jedi* soars, and the redemption scene is a triumph.

Don't argue.

CONSTITUTIONAL EPISODES

Free Speech, Sex Equality, and Same-Sex Marriage as Episodes

The nature of injustice is that we may not always see it in our own times, The generations that wrote and ratified the Bill of Rights and the Fourteenth Amendment did not presume to know the extent of freedom in all of its dimensions, and so they entrusted to future generations a charter protecting the right of all persons to enjoy liberty as we learn its meaning. . . . The Court, like many institutions, has made assumptions defined by the world and time of which it is a part.

—JUSTICE ANTHONY KENNEDY

Star Wars offers an assortment of lessons about fathers and sons, freedom of choice, the possibility of redemption, and even rebellion. But it does not have all that much to say about constitutions, at least not directly. Sure, it's in favor of the separation of powers (empires bad, republics good—mostly), and it opposes squabbling legislatures. It seems to be in favor of human rights: torturers don't come off well, and it's bad to kill innocent people.

But if you're looking to learn about constitutional design, Star Wars probably isn't your best bet. Go see *Hamilton* on Broadway. Or even better, read Gordon Wood's classic, *The Creation of the American Republic, 1776–1787*.

Nonetheless, an understanding of the Star Wars saga does tell us a fair bit about constitutional law—not in terms of its content, but in terms of how it gets created, and the kinds of freedom and constraint that judges have. In short: Constitutional law is full of "I am your father" moments—twists and turns, reversals, unanticipated choices, seeds and kernels that launch whole new narratives. Judges are authors of Episodes, facing a background that they are powerless to change. But they are nonetheless able to exercise a lot of creativity.

At any given moment, the rights that Americans have are different from the rights that they had a few decades before. In the 1940s, government could regulate speech that it deemed to be dangerous. Who would have thought that by 1970, the Constitution would be understood to create a strong principle of freedom of speech, allowing dissenters to say whatever they want? In the late 1950s, sex discrimination was part of the fabric of American life, and the Constitution did not stand in its way. Who would have thought that by 1980, the Constitution would be taken to forbid sex discrimination? In 2000, it was extreme, even radical to say that the Constitution protects the right to same-sex marriage. In 2015, a majority of the Supreme Court ruled that it does exactly that.

But in every one of these cases, the Supreme Court built on an existing narrative. It did not start one. It couldn't! It's not allowed to do that. We're talking here about Episodes XX, and XXX, and XL, not about a brand-new story. At the very least, we can say that the constitutional tale was not planned out in advance, and that the authors of the initial Episodes could not possibly have anticipated what was to come. Someone, or a few people, had to make a choice.

Like Lucas and Abrams, the most powerful judges are creators, and they make choices against the backdrop set of previous Episodes, also known as precedents. They have to figure out how to continue the saga. A brief warning: To explain that conclusion, I'm going to go into some detail here, not least because constitutional law is my day job.

But there's another reason. With the unexpected death of Justice Antonin Scalia in 2016, the United States is in the midst of a heated debate over what it means to interpret the Constitution. The debate becomes especially intense whenever the Supreme Court has a vacancy—but it is always important to have. One reason, I suggest, is that the Court's members, including its newest ones, have the great privilege of deciding on the plots of the new Episodes. They're a lot like George Lucas, J. J. Abrams, and their successors. If the question involves privacy, free speech, sex equality, guns, or the power of the president, the answer will require obedience to the past, but also a judgment about what makes it shine most brightly.

When we disagree about Supreme Court decisions, our arguments are about what makes a new Episode best. I am acutely aware that my old University of Chicago colleague, Justice Scalia, would not agree with me on that point. I liked him and admired him, and I revere his memory—but in law, even Jedi masters sometimes get it wrong.

INFINITE POSSIBILITIES

Let's start by noticing that in the extended universe of Star Wars, and in fan fiction and reactions, we can find a lot of "What if?" speculation. The same kind of speculation makes a lot of sense for law, including those aspects of law that deal with the meaning of the Constitution. Things could easily be otherwise.

One of the most elaborate examples is *Star Wars Infinities*, a series of three graphic novels in which some really small point of divergence jolts the familiar stories in wildly different directions. They're good— and the word "infinities" is well-chosen. What if Luke's efforts to destroy the Death Star had not quite worked? What would have happened? That's the foundation for one of the stories in *Star Wars Infinities*. What if Luke had been killed by that big monster on Hoth? What would Han and Leia have done? That's the foundation for another one.

But insiders have also produced something like *Star Wars Infinities*. George Lucas tried his own VII, VIII, and IX, and Disney rejected them, which meant that the authors of *The Force Awakens*, J. J. Abrams and Lawrence Kasdan, went off on a path of their own. Initially, Abrams worked with a distinguished scriptwriter, Michael Arndt, who produced something, but not quite what he wanted—so Abrams and Kasdan chose their own approach (while continuing to credit Arndt). An intriguing question to which I hope we'll eventually get an answer: What would Lucas's Episodes have looked like?

Even given the backdrop of the first two trilogies, we could imagine many different versions of Episode VII. It might have been set immediately after *Return of the Jedi*, cataloging, in a slow, steady, and upbeat way, the happy restoration of the Republic, the blissful marriage of Han and Leia, and the birth of their four children, with Luke as the friendly uncle with special powers. (Boring!) Or we could imagine an Episode VII of a radically different kind, in which it is revealed that Darth Vader and Emperor Palpatine are still alive, and everything that seemed to happen at the end of *Return of the Jedi* was just Luke's dream. In this version of the Episode, we're right back at the end of *The Empire Strikes Back*. (Terrible! A betrayal of the audience!)

We could imagine another Episode VII, set two years after *Return of the Jedi*, in which Han and Leia break up, because Leia cannot get over her intense, obsessive romantic attraction to her brother. She tries to convince Luke that if it feels so right, it can't be wrong. (Ick.)

Or we could imagine an Episode VII, set five years after the *Return of the Jedi*, in which Luke, having obtained unimaginable power, turns out to be strongly drawn to the Dark Side. (Potentially interesting. It would have a plausible link with the final temptation scenes of *Return*, which it would cast in a whole new light.)

In the real Episode VII, of course, Abrams and Kasdan chose a different approach, essentially replicating the tale of *A New Hope*. In my view, they made it work (though I like Dark Side Luke better). The point is that they began their project meaningfully constrained, but also with countless options. That's true for judges, too. In fact it's true in precisely the same way.

FOLLOWING RULES

How do judges settle constitutional disputes? According to one view, the answer is simple: they read the Constitution. Unless they are playing games, they tell us what it means. On this view, the Constitution is very much like the Journal of the Whills, except that it is real. In the

United States, Republican politicians often argue in favor of a view of this kind. Judges should just follow the law.

In some cases, that's just right. The American Constitution says that to be elected president, you have to be at least thirty-five years old, and that there will be one president rather than two or three, that Congress will consist of both a Senate and a House of Representatives, and that the members of the Supreme Court will have life tenure. Many important things are written down in advance, and there's nothing to do but to adhere to what's written.

At the same time, some of the most important provisions of the Constitution are ambiguous or open-ended. The Constitution uses the word *liberty*. What's that? Does it include the right to use contraceptives? To have an abortion? To marry people of the same sex? To wield a lightsaber? The Constitution protects "the freedom of speech." Does that mean that people have the right to issue threats? To bribe members of the First Order? To use Jedi mind-tricks? To commit perjury? To cry "fire!" in a crowded theater? To recruit people to commit acts of terrorism? The Constitution forbids states from denying any person "the equal protection of the laws." Does that provision ban racial segregation? Does it prohibit discrimination on the basis of sexual orientation? Does it forbid affirmative action programs?

Some people, such as the late Justice Scalia, have a distinctive way of approaching these questions. They want to ask: *What did those provisions mean when they were originally ratified?* On this view, judges can greatly simplify their tasks by going into a kind of time machine, and figuring out what was meant by "We the People" when they ratified constitutional provisions.

If we do that, we might well learn that "freedom of speech" did not include the right to commit perjury, and that "equal protection" had nothing to do with discrimination on the basis of sexual orientation. If the Journal of the Whills existed, perhaps we could figure out its meaning by trying to discover what Lucas actually meant. Scalia and others favor a similar approach to constitutional law.

But the Supreme Court has firmly and repeatedly rejected this approach. One reason is historical: was the Constitution originally understood to set out very specific rules, or instead to set out broad principles whose meaning was supposed to change over time? If the answer is the latter, then Scalia's approach turns out to be self-defeating: the original meaning was that the original meaning would not govern! Some historians think that originalism is inconsistent with the original meaning. It's possible that as a matter of history, those who ratified the founding document rejected originalism and favored something like new Episodes.

Another reason is pragmatic: does it really make sense to interpret the Constitution's broad phrases by asking about their meaning well over two hundred years ago? Is it best to understand "freedom of speech" or "cruel and unusual punishment" by asking what people thought about those phrases in 1789? Those who favor some kind of "living Constitution" insist that that's hardly best, and that judges can legitimately understand constitutional meaning to change as the decades pass. They insist that societies learn over time, and that constitutional meaning can reflect that learning.

If we decide that the meaning of the Constitution changes, then what are judges supposed to do? Should they follow the election returns? Make their own moral judgments? Consult an evolving social consensus? Predict the future? Or just defer to whatever state and national governments have decided, so long as the Constitution does not obviously stand in their way? What, if anything, does Star Wars say about these questions?

LAW AS EPISODES

In his brilliant work on the nature of legal reasoning, the legal theorist Ronald Dworkin provided the arresting metaphor of a chain novel. Assume (my bare-bones example, not Dworkin's) that ten people are charged with producing a novel, with each person asked to produce a specific chapter. Ackbar writes chapter one—say, about Marjorie, apparently a high-level executive with a computer company, who turns out to be seated next to John, who happens to work for the Central Intelligence Agency, on a flight from New York to Berlin. Kylo writes chapter two, and details the conversation between the two. As Kylo develops the plot, they're both divorced, and romantic sparks start to fly between them. Now it is Poe's turn. What will his third chapter look like?

Dworkin urges that if Poe is to be faithful to his task, he will want to make the emerging novel the best that it can be. To do that, he will have to fit what has gone before. If John turns out to be Jabba the Hutt, the story will start to look pretty idiotic (unless Poe is very clever). Simple and abstract though their chapters are, Ackbar and Kylo have imposed real constraints on Poe.

But within the constraints of fit, Poe will have many options, some evidently better than others. If the two characters lose interest in one another and start reading the newspaper, the plot will go nowhere. If Marjorie turns out to be in intense discussions with the National Security Agency, about how to safeguard personal privacy, the story might start to be interesting.

Dworkin claims that the chain novel metaphor tells us a great deal about the nature of interpretation in general and legal interpretation in particular. He's right. Does the Constitution forbid affirmative action programs? Does it require states to recognize same-sex marriage? To answer such questions, judges must investigate their previous decisions, and they must ask what answer puts those decisions in the best light, or makes constitutional law the best that it can be. They must write the next episode. To a significant extent, that is what constitutional law is.

"I DON'T LIKE THAT AND I DON'T BELIEVE THAT," LEGAL STYLE

Recall the debate between Lawrence Kasdan and George Lucas: should a major character die in *Return of the Jedi*? Something like that debate has numerous parallels in constitutional law.

Someone suggests a new development—say, a decision to the effect that the right to privacy includes the right to polygamous marriage. A lawyer gives a reason for that conclusion—say, if privacy means anything, people should be able to marry not just the person they want, but as many people as they want. Someone disagrees ("I don't like that and I don't believe that")—arguing that in terms of past decisions, the right to privacy includes only a right to marry a person, not people. Someone else argues that the protection of polygamous marriage is actually a great idea, because it fits well enough with the precedents, and because it offers the best continuation of the narrative that those precedents started. A skeptic responds that polygamous marriage would destroy the whole point of the marital institution, which the precedents were careful to safeguard. Judges discuss, and they disagree, and then they vote. That's how constitutional law works.

The best historical example is freedom of speech. For many people, including many lawyers, contemporary free speech law is understood as if it emerged from some kind of Journal of the Whills—as if it has been spooled out of something (such as the original understanding of the text, or the basic commitments of James Madison). But that's false; consider the area of commercial advertising.

Until 1976, the Supreme Court had *never* ruled that the First Amendment protects such advertising. In its own "I am your father" moment, the Court decided that indeed it did, with an opinion that claimed continuity with a tradition that it was fundamentally revising: "We begin with several propositions that already are settled or beyond serious dispute. . . . It is precisely this kind of choice, between the dangers of suppressing information, and the dangers of its misuse if it is freely available, that the First Amendment makes for us."

Really? The First Amendment does that? In insisting that it does, the Court cast a new light on, and required a fresh understanding of, everything that had gone before. Until 1976, the First Amendment had not been understood to do any such thing. With commercial advertising, the Court had never held (in nearly two hundred years) that the First Amendment makes the choice between suppressing information and allowing the dangers of its misuse. On the contrary, it had said that public officials are allowed to regulate that kind of speech as it sees fit. The Court was acting a lot like George Lucas, proclaiming continuity at the same time that it was making a bold innovation.

With the recent protection of commercial speech in view, we might venture a radically different understanding of our free speech tradition: It is only *political* speech that has long been at the "core" of that tradition, ensuring that Americans can say what they want about their leaders and their government. But very late in the game (1976!), the Court wrongly added commercial advertising, in a way that compromised and undermined the tradition. The protection of such advertising was a false move, a real-world "I am your cat."

But matters are actually far more complicated than that, and the role of creativity and reversals is much larger. Political speech was not, in fact, protected from the beginning. From the founding until the middle of the twentieth century, there was a fair bit of censorship, and the Constitution was not seen to stand in its way. Public officials were permitted to punish speech that they deemed to be harmful, even if that speech did not create anything like a clear and present danger. As late as 1963, dissenters were at serious risk if the government really wanted to punish them.

True, strong protection is now given to political speech. But that is a creation of a brief (and late, and shining) moment in time, punctuated by several "I am your father" decisions, most prominently in 1964 and 1969. In using the First Amendment as a barrier to the use of state libel law in 1964, the Court set forth a proposition that required a rethinking of all that had come before: "the pall of fear and timidity imposed upon those who would give voice to public criticisms is an atmosphere in which the First Amendment freedoms cannot survive." (Is that true? Probably. But in a parallel world, of the sort that Star Wars did not explore—Star Trek does—the Court did not think so.)

What can be said about the First Amendment can be said about countless domains of constitutional law. Racial segregation is unacceptable, and so are all forms of racial discrimination, but that principle is a product of the 1950s. Sure, the Equal Protection Clause followed the Civil War, but it is a wild stretch, a form of Leia-like

"I've always known," to say that current equal protection law comes right out of the text of the old clause. Religious liberty, as we live it, is a product of the 1960s. The ban on sex discrimination comes from the 1970s. Sharp constraints on affirmative action are a product of the 1990s and 2000s. Maybe most amazing of all: the Supreme Court did not protect individual gun ownership until the twenty-first century—2008, to be precise.

In each of these cases, lawyers and judges work hard to point to a Journal of the Whills, but it just isn't there. Nothing here was inevitable. Without contingent social movements and judgments (to the effect of "that is the greatest thing that we could possibly ever do"), the United States would have seen radically different constitutional understandings, and radically different constitutional rights. If the Supreme Court had taken different routes, they would have seemed no more surprising, and no less foreordained, than what we now observe.

"ORIGINALISM"

We have seen that when he wrote *A New Hope*, Lucas had no idea about what would become major plot developments in *The Empire Strikes Back* and *Return of the Jedi*. It would have been preposterous for him, and for his coauthors and successors, to write further installments with reference to this question: *What was Lucas's original understanding?* With respect to central issues in the Star Wars series, there is no such understanding to consult, and with respect to others, it points in the wrong direction (as Lucas himself concluded). For constitutional law, the problem is immensely compounded because of the long temporal lag (often centuries) between that understanding and current problems, and because of the rise of unanticipated circumstances of multiple kinds (the telephone, television, the Internet, changing roles of women and men).

In any particular period, constitutional law conveys an aura of inevitability, as if the prevailing narrative were planned or foreordained, or a product of some kind of Journal of the Whills. Assuming the role of Jedi, many originalists, evidently concerned to preserve the constitutional status quo, have worked exceedingly hard to demonstrate that wide swaths of current law actually follow from the original understanding, even if that doctrine was established in the 1950s, the 1980s, the 2000s, or last year.

Some claim, for example, that as ratified in the aftermath of the Civil War, the Equal Protection Clause prohibited racial segregation, or sex discrimination, or even discrimination on the basis of sexual orientation. They urge that after the founding, the free speech principle was understood to create something like the broad protection that Americans now enjoy. Also self-portrayed Jedi, many other judges, not tethered to the original understanding, make similar claims, contending that they are speaking neutrally for the purposes of the provisions that they are interpreting, or just spelling out the inner logic of those provisions.

Don't believe them. Whether Jedi or Sith, many authors of constitutional law are a lot like the author of Star Wars, disguising the essential nature of their own creative processes.

IN WHICH ORDER SHOULD YOU WATCH?

In constitutional law, the order of Episodes is set by time. If a case arises in 2019, the Supreme Court can't decide it in 1971. But for Star Wars movies, people have freedom of choice. At the very least, you have a choice if you are introducing children or friends to the series, or if you are coming to them fresh.

Two possibilities are obvious: Release Order (4, 5, 6, 1, 2, 3, 7) and Episode Order (1, 2, 3, 4, 5, 6, 7). For his six films, Lucas strongly recommends Episode Order, and that has the most logic for sure. You can see events as they unfold, and they all make (enough) sense. There's another advantage, and it's more subtle. At the end of *Revenge of the Sith*, Anakin has just become Darth Vader, and so there's massive drama in his entry as the full-fledged Vader—Darth Vader in full—in *A New Hope*. That's its own kind of "whoa."

In my view, though, there is a decisive problem with Episode Order: You lose the surprise and lessen the impact of the best moment in the saga, which is "I am your father." If you've seen 1, 2, and 3, you already know that Darth Vader is Anakin Skywalker and hence Luke's father. You also lose the various mysteries and coolnesses of *A New Hope*. (Who's Obi-Wan? What's the Force?) So with Episode Order, the two best movies become a lot less interesting. As between the major options, Release Order is best.

Some people have come up with creative alternatives. How about 4, 5, 1, 2, 3, 6, 7? Think about it for a minute. That approach has the advantage of giving you "I am your father," and of starting with the mysteries of the two best, while treating the prequels as kind of a flashback (as you're also focused on the cliffhanger ending of 5). Then you get to wrap everything up with the real finale, and the best, before the third trilogy starts. Not a bad idea at all.

A subversive tweak is called Machete Order: 4, 5, 2, 3, 6, 7. With that approach, you drop *A Phantom Menace*, and you don't lose a whole lot. You miss Anakin as a little kid, which isn't so terrible, and while *A Phantom Menace* has its moments, it really isn't necessary for the arc of the plot. Machete Order is a pretty good idea. (But I have a fondness for *A Phantom Menace*, even if it is the worst.)

There are of course other possibilities. How about Random Order: 6, 4, 3, 1, 7, 2? Maybe if you've seen them all before, or if you're inebriated? What about Reverse Episode Order: 7, 6, 5, 4, 3, 2, 1? Maybe if you want to be challenged by a kind of brainteaser, or if you want to save the worst for last?

The verdict? Release Order. That's how you do it.

THE FORCE AND THE MONOMYTH

Of Magic, God, and Humanity's Very Favorite Tale

Throughout the inhabited world, in all times and under every circumstance, myths of men have flourished, and they have been the living inspiration of whatever else may have appeared out of the activities of the human body and mind.

—JOSEPH CAMPBELL

It's a bit wild to think that a human being, or someone who looks like Yoda, could levitate objects. Isn't it? But if you really can master the Force, you can do lots of amazing things.

Here's a catalog:

Control weak minds

Extract information from minds (might be limited to the Dark Side)

Sense where things are, without actually seeing them

See the future, kind of

Release energy from fingertips to hurt/kill enemies (might be limited to the Dark Side)

Race pods really well (fast and accurately, without crashing or getting killed)

Sense people's location, especially if they are related to you or if they too have the Force

Make people choke

Lift people up and throw them around

Jump and twist around surprisingly fast

Levitate objects, especially lightsabers

Fight spectacularly well with lightsabers

Feel disturbances in the Force, apparently triggered by big events (such as planets blowing up)

Come back after dying (might be limited to the Light Side)

Fire missiles that blow up Death Stars into small cracks in Death Stars (note: the best pilot in the galaxy, Poe Dameron, can do that, too, even though he might not have the Force)

Outside of the Star Wars universe, can anyone do those things? Advertisers and politicians are certainly able to control weak minds.

("These are not the droids you're looking for" might be their motto.) In fact there is a whole literature on what we might count as Jedi mind-tricks. In psychology and behavioral economics, people have shown that if you just describe options in a certain way, or make some features of a situation salient, you can get people to do and even to see what you want. You don't have to be a Jedi to manipulate people's attention.

A small example: If you tell people that out of a group of patients who have a certain operation, 90 percent are alive after ten years, there's a good chance that they'll choose to have the operation. If you tell them that out of the same group, 10 percent are dead after ten years, it's more likely that they'll decline. But of course "90 percent are alive" means exactly the same thing as "10 percent are dead." Describing a problem is a way of "framing" it, and a frame can be an effective mind-trick.

In 2015, two Nobel Prize winners, George Akerlof and Robert Shiller, published an important book with an unusual title: *Phishing for Phools*. The basic idea is that some people, called phishermen, are able to trick other people, called phools. Phishermen may sell credit cards, mortgages, cigarettes, alcohol, or unhealthy foods. They manipulate and deceive people (phools). While they're not exactly Sith, they're certainly not Jedi Knights. But they have the power to control what other people see. Obi-Wan nailed the point: "Your eyes can deceive you. Don't trust them."

Here's one way to look at it. Behavioral scientists, most famously Daniel Kahneman in *Thinking, Fast and Slow*, make a distinction between "fast thinking," associated with the brain's System 1, and "slow thinking," associated with the brain's System 2. System 1 is rapid, intuitive, and sometimes emotional; it sees a big dog and gets scared, or it sees a Star Wars toy and needs to buy it immediately. System 2 is more reflective and deliberative; it sees a big dog and knows it's probably friendly, and with respect to toys, it's willing to say, "I already have eighty-one Star Wars toys, and that's enough." Jedi Knights are able to appeal to people's System 1, and to move them in their preferred directions.

A few months ago, I explained the immediately preceding paragraph to my son Declan. He likes toys, a lot, and whenever we pass a toy store, he has a hard time resisting them. About a week after hearing the explanation, he passed a toy store and asked me, "Daddy, do I even have a System 2?"

Within psychology, an important discussion is Robert Cialdini's *Influence*, which offers six Jedi-type tricks. One of these is *reciprocity*. People like to return favors, and if you give someone something (a discount, a little cash, a token), you'll probably get something back. Another principle is *social proof*: If a lot of people seem to think something, or to do something, others will be inclined to think it or do it, too. (A good way to change behavior is to tell people that other people are now thinking or doing what you want them to think or do.) Another is *scarcity:* people find things more attractive when they seem hard to get, or sharply limited in availability.

RECOGNIZING PATTERNS

Okay, maybe that's not really Jedi material. But great athletes seem to see things when they can't possibly see them; they are sometimes said to have "eyes in back of their head." Consider Tom Brady, the quarterback for the New England Patriots, who seems able to sense the presence of linebackers and defensive linemen when they aren't within his range of vision. In basketball, many players have mastered "no-look passes"; Magic Johnson got his nickname for a reason. In baseball, pitchers display extraordinary reaction time in catching balls hit directly at them.

In my spare time, I play the sport of squash—a bit like tennis, but indoors, with a little ball that can go up to 170 miles per hour. I have had the privilege of practicing with some of the world's best; their powers of anticipation are uncanny. They're here, and then they're there, and that's exactly where the ball is. A great squash player I know had a period of near-blindness in one eye. It didn't much affect her, because she knew where the ball was going. Some athletic skills are all the more impressive when part of a different culture; take a look sometime at the game of competitive *chinlone* (cane ball), played in Burma, and see if it doesn't seem like something out of Star Wars. And of course, a lot of people are astonishingly good at steering fast-moving vehicles (including cars and airplanes).

By the way, there is an explanation for the athlete's apparent ability to see things before they're there. It involves *pattern recognition*. That might well be the essential skill of the Jedi (and of the Sith). If you feel the Force, you can see patterns where other people see a mere blur. That's how you know just what to do. That's why you don't need a computer to shoot into a small space on a Death Star. That's why you can "trust your feelings." Your feelings are your intuitions, and they've been honed to perceive things that others cannot.

That's real, not just Jedi stuff. In baseball, basketball, football, tennis, and squash, for example, accomplished athletes are able, essentially immediately, to perceive familiar patterns, where the rest of us see chaos or noise. They're like great chess players who can take a rapid glance at a chessboard and know exactly what is going to happen. ("White mates in four.") There's no magic here, and there's no need to feel the power of the Force. It's a product of practice and repetition, producing instantaneous recognition of multiple aspects of a situation. As a result of practice, great athletes have keenly educated System 1, and that seems like Jedi stuff. Speaking to Han, Obi-Wan came close to making the point: "In my experience there is no such thing as luck, my young friend—only highly favorable adjustments of multiple factors to incline events in one's favor."

If this seems puzzling, think about driving a car. At first, it's completely confusing. How hard should you push on the accelerator? What do you do when a car is turning into your lane? You have to make a lot of decisions every minute. But after a while, that's the easiest thing in the world. The reason is that the patterns are so familiar. You understand them instantly.

FOR SOME PEOPLE, THE FUTURE ISN'T ALL THAT CLOUDED

No one can see the future, you might think, but as Philip Tetlock and Dan Gardner have shown, some people really are "superforecasters"; they have an uncanny ability to predict what's going to happen. Notably, these particular Jedi tend to agree with statements like these, which might be taken as part of a real-world Jedi Code:

1. Nothing is inevitable.

2. Even major events like World War II or 9/11 could have turned out very differently.

3. People should take into consideration evidence that goes against their beliefs.

4. It is more useful to pay attention to those who disagree with you than to pay attention to those who agree.

By the way, they disagree with these:

1. Randomness is rarely a significant event in our personal lives.

2. Intuition is the best guide in making decisions. (Sorry, Obi-Wan; "trust your feelings" is not always the best advice.)

3. It is important to persevere in your beliefs even when evidence is brought to bear against them.

To be sure, the superforecasters don't feel anything like the Force. They are able to think well about probabilities, breaking up the components of possible future paths in order to get a sense of which really is most likely. They are willing to agree with the claim that "clouded, the future is"—but they are exceptionally good at seeing where the clouds come from, and how big they are.

At the same time, some of the Jedi's powers do seem beyond standard human capacity, because the Force gives people supernatural abilities. In Obi-Wan's famous and mysterious words, it is "an energy field created by all living things. It surrounds us, penetrates us, and binds the galaxy together." Surely no one can levitate objects or come back after the dead. Whether or not there's a Force, most people can't use it.

Or can they?

THE GORILLA THAT DISAPPEARED

A few years ago, I found myself in a group of about thirty people, watching a big video screen in a college classroom. A friend of mine, named Richard, asked us to watch a short video, in which people were throwing basketballs to one another. He also gave us a little task: To count the number of passes. The task wasn't easy. After about 45 seconds, I lost track—but kept trying. The video ended after about 82 seconds, at which point Richard asked us how many passes had been made. Few of us were able to answer correctly.

Then he also asked, "Oh, did you see the gorilla?" All of us laughed. Well, all but one, who raised his hand and said, "I did."

At the time, I was sure that the guy who raised his hand was playing some kind of joke on us. But then Richard said, calmly: "Watch the movie again." Sure enough, a gorilla comes onto the screen, about halfway through, and it beats its chest, and then it moves on. It's as plain as day, and it's on-screen for about nine seconds. There's nothing hidden or obscure about it. And yet almost all of us missed it.

I recently showed the film to my grown daughter Ellyn, having told her what I have just said here. After seeing it, she asked, "Where was the gorilla?" Apparently you might miss the gorilla even if you've been told that there's a gorilla!

Most groups do a bit better than did my own; usually about half of the participants don't see the gorilla. But half is a lot. As Christopher Chabris and Daniel Simons, the designers of the experiment, report a frequent conversation with their subjects:

Q: Did you notice anything unusual while you were doing the counting task?

A: No.

Q: Did you notice anything other than the players?

A: Well, there were some elevators, and S's painted on the wall. I don't know what the S's were there for.

Q: Did you notice *anyone* other than the players?

A: No.

Q: Did you notice a gorilla?

A: A what?!?

The invisible gorilla experiment is profoundly interesting, because it tells us something important about the nature and limits of human attention, and hence about how to do Jedi mind-tricks. Human beings have limited mental "bandwidth," and so we focus on some, but not all, aspects of what we are capable of seeing. The technical term is *inattentional blindness,* and that's what Obi-Wan, Luke, and Rey exploit (kind of). Because of our limited bandwidth, we miss things that are right in front of our eyes—and we can be manipulated. Moreover, we are unaware of that fact. As Chabris and Simons put it, "We experience far less of our visual world than we think we do." As a result, you don't have to be a Jedi or a Sith to turn people's attention where you want.

Professional magicians know this well. Apollo Robbins, pickpocket and magician, emphasizes, "We have only so many mental dollars that we get to spend." Once they're consumed, "the victim has no more left to focus on what is really happening. Presto! The wallet is gone." Robbins uses jokes as a means of eating up "some of the brain's bandwidth." He argues that the idea is to engage the brain's "two security guards." The idea is to get the two talking to each other about what to watch out for, making thievery easier to conduct while the metaphorical guards are distracted. Magicians are terrific at misdirection, which means that they lead the spectator to focus on something other than the cause of the magic effect.

As Robbins puts it, "Attention is like water. It flows. It's liquid. You create channels to divert it, and you hope that it flows the right way." Couldn't Yoda have said just that, more or less? (Like water, attention flows. Yes.)

A TWENTIETH-CENTURY JEDI

But are there really Force Ghosts? Did the twentieth century see a Jedi Knight, able to conjure people from beyond the grave? More than one? In a captivating book, *The Witch of Lime Street*, David Jaher tells us a great deal about how magic works, and perhaps about something like the Force as well.

As Jaher explains, some of the world's greatest thinkers were convinced, in the 1920s, that people could speak to the dead. Sir Arthur Conan Doyle created Sherlock Holmes, the canonical detective, who could always see through fakery and artifice. Having lost a son in the Great War, Doyle was also a convinced Spiritualist who thought death "rather an unnecessary thing." In his popular 1918 book, *The New Revelation*, he argued vigorously on behalf of spiritualism. His dedication: "To all the brave men and women, humble or learned, who have the moral courage during seventy years to face ridicule or worldly disadvantage in order to testify to an all-important truth." From 1919 to 1930, Doyle wrote twelve more books on the same subject.

In the 1920s, as now, *Scientific American* was a highly respected publication, dedicated to the dissemination of research findings. In 1922, Doyle challenged the magazine and its editor-in-chief, Orson

Munn, to undertake a serious investigation of psychic phenomena. James Malcolm Bird, an editor there (and previously a mathematics professor at Columbia University), was intrigued. In November the magazine established a highly publicized contest, with a prize of five thousand dollars for anyone who could produce conclusive evidence of Jedi-like "physical manifestations"—as, for example, by making objects fly around the room. The magazine soberly announced that as of yet, it was "unable to reach a definite conclusion as to the validity of psychic claims."

All of the initial candidates failed that test; the committee saw through them. In the meantime, a woman named Mina Crandon was garnering international attention. One friend, speaking for many, described her as a "very very beautiful girl" and "probably the most utterly charming woman I have ever known." She also appeared able to levitate objects (tables and chairs) and to speak to the dead, above all her brother, Walter.

In London, she performed in front of several investigators, appearing to make a table rise and float. She and her husband became friendly with Doyle, who swore to "the truth and range of her powers." Bird invited her to enter the magazine's contest. Accepting the challenge, she moved objects, producing noises in various places, and channeled Walter. In the July 1924 issue of *Scientific American,* Bird wrote about her, protecting her privacy with the name "Margery." He said that "the initial probability of genuineness [is] much greater than in any previous case which the Committee has handled." Bird's article was widely discussed. A headline in the *New York Times* read, "Margery Passes All Psychic Tests." The *Boston Herald* exclaimed, "Four of Five Men Chosen to Bestow Award Sure She Is 100 P.C. Genuine."

It took Harry Houdini, the acclaimed magician, to expose her.

Just like Han to Luke, Houdini insisted that "it's all a lot of simple tricks and nonsense." Observing her closely on several occasions, Houdini began to figure out exactly how she produced some of her most impressive effects. With evident admiration, he reported, Crandon had produced "the 'slickest' ruse I have ever detected, and it has converted all skeptics." In November 1924, he wrote a lengthy pamphlet, complete with highly detailed drawings of the séances, with which he specified exactly how Crandon was able, in the dark, to maneuver her legs, head, and shoulders to produce the various effects.

But Crandon's numerous defenders were unconvinced. They portrayed Houdini as implacably close-minded, himself a cheat. Doyle denounced Houdini as prejudiced and dishonest; the denunciation destroyed their friendship. Even years later, Doyle proclaimed, "I am perfectly certain that it was an exposure, not of Margery but of Houdini."

How was Margery able to dupe so many people, including some of the great thinkers of the day? The answer has a lot to do with a Jedi mind-trick, in the form of an extraordinary ability to manipulate people's attention. Consider a little tale from one of Margery's investigators, Princeton psychologist Henry C. McComas, who described her supernatural feats to Houdini with great wonder, insisting that he saw every one of them with his own eyes. McComas reported that for the rest of his life, he would not forget the scorn with which Houdini greeted those words. "You say, you *saw*. Why, you didn't see anything. What do you see now?" At that point, Houdini slapped a half dollar between his palms, and it promptly disappeared.

His great adversary never confessed. In her very last days, a researcher suggested to a failing Mina Crandon, widowed for two years, that she would die happier if she finally did so, and let the world know about her methods. To his surprise, her old twinkle of merriment returned to her eyes. She laughed softly and offered her answer: "Why don't you guess?"

That's the best answer, of course, but there's little doubt that Margery used a lot of invisible gorillas. Focusing people's attention in some places, and diverting their attention from others, she was able to make them see exactly what she wanted them to see. In short: "These are not the droids you are looking for."

JEDI NUDGES, SITH NUDGES

With my friend and coauthor, the great economist Richard Thaler, I have explored the idea of "nudges"—interventions by private and public institutions that fully preserve freedom of choice, but that also steer people in certain directions. A GPS device is a nudge: it tells you the best way to get where you want to go. A reminder is a nudge ("don't forget your lightsaber, dear"). So is a warning ("there's an asteroid on the right") or simple information. If you tell people about the social norm ("most people avoid the Dark Side"), you're nudging, and the same is true if you "frame" a situation so as to promote a certain kind of conduct ("ninety percent of people who try to learn how to use the Force get it after three months").

All over the world, national governments have shown a keen interest in nudging. In 2010, the United Kingdom created a Behavioural Insights Team. In 2014, the United States followed with its own Social and Behavioral Sciences Team. Australia, Germany, and the Netherlands have created similar teams. They don't use Jedi mind-tricks, but they do enlist understandings of behavioral economics and human psychology to try to make government work better.

We might make a distinction here between two kinds of nudges: those that are open and transparent and those that seem more covert. A calorie label, or a warning about the risks of smoking, is completely open. Nothing is hidden or obscure. But if healthy foods are placed at eye level, and if less healthy ones are more difficult to see, some people think that there is a risk of manipulation. Subliminal advertising is far worse, because people are not aware that they are being influenced.

Both Jedi and Sith are capable of nudging—transparently and covertly. In fact they do some of both. And while the Force enables them to work on weak minds, they seem to have a kind of ethical constraint: they usually want people to choose, and they want people to choose *freely*. We do not know, for sure, whether Obi-Wan, Yoda, Vader, or the Emperor could work on Luke's mind without his consent. But we do know that they want him to choose a particular way. I have noted that in an echo of the tale of Dr. Faust, the Dark Side seeks Luke's soul, and he has to give that up by his own free will.

Whenever Thaler signs our book *Nudge*, he is careful to write: "Nudge for Good." He does that because an understanding of human psychology opens up opportunities for hurting people—for using their own intuitions against them. Private companies sometimes do exactly that. So do unsavory politicians. Phishermen aren't exactly Sith, but they nudge, and not for good.

OLD MYTH, NEW WAY

But the Force is not merely about human psychology, behavioral biases, or even magic. It is far murkier and more mysterious than that. Above all, it involves a "leap of faith." Qui-Gon insisted that "[t]he ways of the Living Force are beyond our understanding."

Undoubtedly so, but the ways of George Lucas are pretty transparent, at least here. He was and remains intensely interested in religions, and he sought to convey something spiritual. When he was just eight years old, he asked his mother, "If there's only one God, why are there so many religions?" He's been fascinated by that question ever since. In writing Star Wars, he said, "I wanted a concept of religion based on the premise that there is a God and there is good and evil. . . . I believe in God and I believe in right and wrong."

Star Wars self-consciously borrows from a variety of religious traditions. Lucas thinks that in an important sense, all of them are essentially the same. He is clear about that, insisting that in doing that borrowing, he "is telling an old myth in a new way." We have seen that he was immensely influenced by Joseph Campbell, his "last mentor," who claimed that many myths, and many religions, were rooted in a single narrative, a product of the human unconscious. Campbell can be taken to have given a kind of answer to eight-year-old Lucas: there is one God, and all religions worship Him. Campbell argued that apparently disparate myths drew from, or were, the "monomyth," which has identifiable features.

In brief: A hero is called to some kind of adventure. (Perhaps by circumstances, perhaps by someone in distress.) Initially he declines the call, pointing to his fears, his habits, and what he can't do. But eventually, he feels compelled to accept the call and leaves his home. Encountering serious danger, he needs, and obtains, supernatural aid, often from a small, old, or wizened man or woman. (Think Obi-Wan or Yoda.) He is initiated through various trials, some of them life-threatening, but he manages to survive. Then he faces some kind of evil temptation, perhaps from a satanic figure, whom he resists (with severe difficulty). At that stage he has a reconciliation with his father—and becomes godlike, a religious figure (the apotheosis). Defeating the most dangerous enemies, he returns home to general acclaim.

That is, of course, a summary of many myths and many religious traditions; it also captures countless books, television shows, and movies in popular culture. (The Matrix, Batman, Spider-Man, Jessica Jones, and Harry Potter are just five examples; many comic books, and the movies based on them, have a similar plot.) In a nutshell, it's Luke's journey in the first trilogy. In Lucas's words, "When I did Star Wars, I self-consciously set about to recreate myths and the classic mythological motifs." The Hero's Journey also captures much of Anakin's in the prequels—with the terrific twist that Anakin becomes a monster, not a savior. But as it turns out, he's the ultimate savior, the Chosen One who restores balance to the Force, and so his journey nicely fits the standard pattern if the six episodes are taken as a whole. Seeing the first trilogy for the first time, Campbell was inspired: "You know, I thought real art stopped with Picasso, Joyce, and Mann. Now I know it hasn't."

As Lucas put it, "With Star Wars, it was the religion—everything was so taken and put into a form that was easy for everybody to accept so it didn't fall into a contemporary mode where you could argue about it. It went everywhere in the world." The enduring triumph of Star Wars is that it takes a familiar tale, built into disparate cultures and psyches, sets it in a wholly unfamiliar setting, makes it effervescent and fresh, and gives it a series of emotionally daring twists, thus allowing a series of kids' movies to touch the human heart. Our modern myth is both a spiritual quest and a psychodrama, insisting that redemption is always possible, that anyone can be forgiven, and that freedom is never an illusion.

OUR MYTH, OURSELVES

Why Star Wars Gets to Us

Born as a knockoff of the old Flash Gordon series, Star Wars is a little like a childhood memory, and it's a little like a first kiss, and it's a little like a Christmas present. It's a little like air. Star Wars is here to stay.

Timing is everything, and luck matters. In 1977, the time was certainly right for an upbeat tale about heroes, hermits, droids, and lightsabers. After assassinations, turmoil, and malaise, the United States needed a big lift, and *A New Hope* gave it one. In 2015, the relaunch greatly benefited from the era's evident taste for nostalgia (sequels and more sequels) and its compelling need for good news. The familiar cast of characters could link people with their own youth and with their parents, alive or dead—and with their children, too. After the Great Recession, and in the midst of terrorist threats, Rey, Finn, Poe, and the Resistance were irresistible. (Han Solo, too, even if he died.)

People also tend to like things that other people like. Whenever there's a big fuss, most of us want to know what it is all about. There's a deep human desire for common knowledge and common experiences.

Nations need celebrations and events that diverse people can share; holidays, movies, television shows, and sporting events provide them. The release of a new Star Wars movie is a national celebration.

It might not even matter all that much if it is good! If a new Episode connects you with millions of people in your city, or with people all over the country or even the world, well, that can fill the human heart. In a fragmented world, full of niches and echo chambers, Star Wars provides much-needed connective tissue. You might be young or you might be old, you might be a Democrat or you might be a Republican, but you can have a good argument over whether Han shot first, or whether the prequels are underrated, or the real motivations of Rey and Kylo.

Star Wars has a lot to say about empires and republics, and it draws directly on the fall of Rome and the rise of Nazism. Its simple, stylized claims about what's wrong with empires resonate in many nations. But it's not didactic. Is it feminist? (Kind of.) Is it about Christianity? (Yes.) Does it embrace Buddhism? (It tries, at times, but nope, not at all, anything but.) You can interpret it in countless ways; it invites disagreement and obsessions.

The Force remains mysterious, but each of us is able to recognize the Light Side, and also the Dark. Star Wars is keenly aware that the human heart houses both. Lucas was not of the Devil's Party, and Abrams isn't, either, but they are alert to its appeal. Star Wars might be a bit too earnest for William Blake, who spent a lot of time with the Dark Side. ("Energy is Eternal Delight.") But he would have appreciated it.

Star Wars portrays, and triggers, some of the deepest feelings of children for their parents, and parents for their children. It captures the overwhelming intensity of those feelings—and their ambivalence as well. When a father or his son witnesses Vader save Luke, or Kylo kill Han, we're going back to Greek tragedy, to Freud, and to human fundamentals. Joseph Campbell, Lucas's Yoda, pointed to people's need to "feel the rapture of being alive, that's what it's all finally about, and that's what these clues help us to find within ourselves." Star Wars contains those clues.

Star Wars is a space opera, but its best moments are surprisingly intimate. They don't involve ships, explosions, or strange creatures, or republics and rebellions. In those moments, one human being sees, and insists on, the good in another, even in the aftermath of the most terrible acts. It's face-to-face. Even more than mercy, forgiveness "is twice blest," because "it blesseth him that gives and him that takes." With a little luck, and a resolution to love oneself despite everything, an insistence on forgiveness can produce redemption, possibly in the form of acts of spectacular courage.

For all its talk of destiny, Star Wars insists on freedom of choice. That's its largest lesson. Through acts of personal agency, people can alter the seemingly inevitable movements of history. On a small scale or a large one, they can set things right. Farm boys can decide to go to Aldaraan. Self-interested smugglers can choose to come back, and with a single shot, they can rescue their buddies. ("YAHOO! . . . You're all clear, kid, now let's blow this thing and go home!") Seeing blood on their helmets, Stormtroopers can elect to leave the First Order and help a prisoner with mischief in his eye, who turns out to be the best pilot in the galaxy. Scavengers can choose to save a little droid called BB-8, and find out that the most famous lightsaber in the galaxy belongs in their own hands.

Star Wars is primal, and it's a fairy tale, but it's no mere retelling of Campbell's monomyth. It's far more superficial, and it's much deeper.

It's Flash Gordon, and it's a western, and it's a comic book. It claims to honor destiny, but its real topic is the fork in the road and the decision you make on the spot. With a holler and a whoop, it turns out to be all-American. Still, it manages to be universal, focusing as it does on the most essential feature of the human condition: freedom of choice amid a clouded future.

Star Wars pays due tribute to the importance of distance and serene detachment. But its rebel heart embraces intense attachments to particular people, even in the face of lightning bolts from the Emperor himself. At the decisive moment, children save their parents. They are grown. They announce their choice: "I am a Jedi, like my father before me."

BIBLIOGRAPHICAL NOTE

There are so many books and articles on Star Wars, and so many of them are interesting and good, that it seems uncharitable to single out just a few. But I have learned most from Michael Kaminski, *The Secret History of Star Wars* (Kingston, Ontario: Legacy Books Press, 2008), which is also a ton of fun; Chris Taylor, *How Star Wars Conquered the Universe*, revised and expanded edition (New York: Basic Books, 2015), which makes for terrific one-stop shopping into the topic; and from the comprehensive, fabulous treatments by J. W. Rinzler, *The Making of Star Wars: The Definitive Story Behind the Original Film* (New York: Del Rey Books, 2007); *The Making of Star Wars: The Empire Strikes Back* (New York: Del Rey Books, 2010), and *The Making of Star Wars: Return of the Jedi* (New York: Del Rey Books, 2013). George Lucas also gives excellent, illuminating interviews, and many are collected in Sally Kline, ed., *George Lucas: Interviews* (Jackson: University Press of Mississippi, 1999).

For some of the discussion here, I have drawn on social science research that does not (shockingly) engage Star Wars. A brilliant treatment of contingency, history, and social influences is Duncan Watts, *Everything Is Obvious* (New York: Crown Business, 2011); Watts has inspired much of my discussion here. Michael Chwe, *Rational Ritual* (Princeton, NJ: Princeton University Press, 2001), has a lot to say about collective experiences; it is short but profound.

On behavioral science, two terrific sources are Daniel Kahneman, *Thinking, Fast and Slow* (New York: Farrar, Straus & Giroux, 2011), and Richard H. Thaler, *Misbehaving* (New York: Norton, 2015). On informational cascades, the original analysis is Sushil Bikhchandani, David Hirshleifer, and Ivo Welch, "A Theory of Fads, Fashion, Custom, and Cultural Change as Informational Cascades," 100 *Journal of Political Economy* 992 (1992). Group polarization is explored in Cass R. Sunstein, *Going to Extremes* (Oxford and New York: Oxford University Press, 1999). On constitutional law as Episodes, Ronald Dworkin, *Law's Empire* (Cambridge, MA: Belknap Press, 1985), remains the defining treatment.

ACKNOWLEDGMENTS

I didn't plan to write this book, and if you told me that I was going to, I wouldn't have believed you. The project started less than a year ago, when my wife and I were having dinner at the home of two good friends, Jenna Lyons and Courtney Crangi. As the evening wound down, Courtney casually pointed to an old compact disk containing *A New Hope*. She said that I should borrow it and show it to my son Declan, then five years old.

I hadn't seen the movie for decades and had no particular desire to see it again. Declan was interested in baseball, not spaceships, and he was a bit young for droids, blasters, and Lord Vader. So showing him the movie seemed pretty doomed. But on a lark (and to be polite to Courtney), I gave it a try. Of course he loved it. I did, too.

After seeing *A New Hope*, we promptly saw the five others (though just a part of *Revenge of the Sith*, which is pretty intense). I started to get a bit obsessed. Thanks, Courtney.

For decades, I have invited (okay, begged) law students to help me on research projects, involving such subjects as the Administrative Procedure Act, regulatory reform, the value of a statistical life, and default rules in environmental law. I have always been lucky enough to get a good response, but for this book, the response was unparalleled. In fact it was overwhelming. Special thanks to Declan Conroy, Lauren Ross, and Christopher Young—Jedi Knights all.

Heartfelt thanks as well to Jacob Gersen, David Jaher, Martha Nussbaum, L. A. Paul, Richard Thaler, and Adrian Vermeule for comments on all or part of the manuscript. Particular thanks to Vermeule not only for numerous discussions but also for publishing a review-essay on Star Wars in the *New Rambler*, which he edits; this book grew from that seed. (The essay can be found at http://newramblerreview.com/book-reviews/fiction-literature/how-star-wars-illuminates-constitutional-law-and-authorship.)

Thanks to my terrific agent, Sarah Chalfant, for her support, guidance, and enthusiasm. For this law professor, Star Wars was not exactly a likely topic, and I was genuinely surprised, and remain more than grateful, that Sarah encouraged me to proceed. I am also grateful to members of a reading group I taught at Harvard Law School in the fall of 2015, on the topic of contingency and serendipity. The course wasn't about Star Wars, but the subject did (ummm) come up. I thank as well Tom Pitoniak for an excellent, careful copyedit.

My wife, Samantha Power, is not a huge Star Wars fan, but she saw *The Force Awakens* with me, and she actually liked it. She has also been generous enough to tolerate countless discussions of Luke, Leia, Obi-Wan, Darth Vader, and all the rest—and to keep her good cheer while Declan, Rian, and I stared at Episodes on the computer. (If she felt left out, she didn't show it.) Amazingly, she's shared my enthusiasm for this project. Astonishingly, she read an early draft of this book, in full, and she made large-scale suggestions about how to structure it, and also numerous page-by-page edits, which reoriented and greatly improved the manuscript. The Force runs very strong in her family (Anakin-level midi-chlorians, no doubt); I am truly blessed to be part of it.

Julia Cheiffetz was, and is, the best editor ever. She's brilliantly creative, and she's tremendous fun, and she has vision. Actually she's a bit like George Lucas, in the sense that her standards are really high, and she won't just settle. I am keenly aware that this book is not nearly as good as Julia deserves—not close—but her efforts made it a ton better than it would otherwise be. For whatever works here, she's been my copilot.

My father's favorite place on earth, I think, was Marblehead, Massachusetts. He adored Preston Beach, and fishing, and tennis, and his children, and soft-serve ice cream, which amazed him. In my entire life, I never saw him angry (not even once). He died in his young sixties; he didn't live long enough to meet my three children. With his big, strong shoulders and unfailing, broad smile, he never got old. He had no Darth Vader in him, and no Kylo Ren, and only a bit of Obi-Wan—but plenty of Han Solo. (He was a great flirt.) When I was a child, he showed me his World War II medals—and while he never gave me a lightsaber, I have those medals now. Thanks, Dad.

NOTES

INTRODUCTION: LEARNING FROM STAR WARS

1   **"All the gods"**: Joseph Campbell, *The Power of Myth* (Anchor, 1991), 46.

1   **As of early 2016**: See "Star Wars Total Franchise Revenue," Statistic Brain Research Institute, http://www.statisticbrain.com/star-wars-total-franchise-revenue/ (last visited February 14, 2016).

6   **"It's the biggest adventure"**: Adam Rogers, *"Star Wars'* Greatest Screenwriter Wrote All Your Other Favorite Movies Too," *Wired,* November 18, 2015, http://www.wired.com/2015/11/lawrence-kasdan-qa/.

EPISODE I: I AM YOUR FATHER:
The Heroic Journey of George Lucas

10   **"Scientists and writers, for example"**: Daniel Kahneman and Amos Tversky, "The Availability Bias," in *Judgment Under Uncertainty: Heuristics and Biases* (Daniel Kahneman et al. eds., 1982), 415.

11   **"You have to remember"**: Anwar Brett, "Interview with George Lucas," *BBC,* http://www.bbc.co.uk/films/2005/05/18/george_lucas_star_wars_episode_iii_interview.shtml (archived on September 24, 2014).

11   **"The Star Wars series started out"**: Chris Taylor, *How Star Wars Conquered the Universe*, revised and expanded edition (New York: Basic Books, 2015), 115.

12   **"From the outset I conceived Star Wars"**: George Lucas, "Introduction," in Donald F. Glut, *Star Wars V: The Empire Strikes Back* (New York: Del Rey Books, 1980).

12   **"a western movie set in outer space"**: J. C. Macek III, "Abandoned 'Star Wars' Plot Points Episode IV: A Family that Slays Together Strays Apart," *Pop Matters*, June 22, 2015, http://www.popmatters.com/feature/194139-abandoned-star-wars-plot-points-episode-iv-the-family-that-slays-tog/.

13   **"*The Star Wars* is a mixture"**: Taylor, *How Star Wars Conquered the Universe*, 111.

13   **"I wanted to do Flash Gordon"**: Sally Kline, ed., *George Lucas: Interviews* (Jackson: University Press of Mississippi, 1999), 219.

13   **The first synopsis**: Jan Helander, "The Development of Star Wars as Seen Through the Scripts by George Lucas" (1997), available at http://hem.bredband.net/wookiee/development/.

13   **"I wrote the first version"**: Kline, ed., *George Lucas: Interviews*, 57 (hereinafter referred to as *Interviews*).

13   **"Our plan was to do Star Wars"**: "Did 'Star Wars' Become a Toy Story? Producer Gary Kurtz Looks Back," Hero Complex, August 12, 2010, http://herocomplex.latimes.com/movies/star-wars-was-born-a-long-time-ago-but-not-all-that-far-far-away-in-1972-filmmakers-george-lucas-and-gary-kurtz-wer/.

13   **Lucas produced a list**: J. W. Rinzler, *The Making of Star Wars* (New York: Del Rey Books, 2007), 8.

14   **There was an ice planet**: Michael Kaminski, *The Secret History of Star Wars* (Kingston, Ontario: Legacy Books Press, 2008), 45.

14   **"would hurtle and tumble"**: Taylor, *How Star Wars Conquered the Universe*, 103.

14   **"This is the story of Mace Windy"**: Rinzler, *The Making of Star Wars*, 8.

14   **C. J. stood for**: Ibid.

15   **"Warlord to the Chairman"**: Ibid.

15   **In this brief document**: Taylor, *How Star Wars Conquered the Universe*, 103.

15   **"The two terrified, bickering bureaucrats"**: Kaminski, *The Secret History of Star Wars*, 51.

16   **"It is the thirty-third century":** Ibid., 52.

16   **It is a myth:** The best discussion is Michael Kaminski, *The Secret History of Star Wars* (Kingston, Ontario: Legacy Books Press, 2008), 469–86. Note, however, that Lucas himself said: " 'Darth' is a variation of dark, and 'Vader' is a variation of father. So it's basically Dark Father." Gavin Edwards, "George Lucas and the Cult of Darth Vader," *Rolling Stone,* June 2, 2005, http://www.rollingstone.com/movies/news/george-lucas-and-the-cult-of-darth-vader-20050602.

17   **They converse, in a way that's also a bit familiar:** Taylor, *How Star Wars Conquered the Universe,* 113.

17   **"perhaps the most curious":** Kaminski, *The Secret History of Star Wars,* 447.

18   **"about Ben and Luke's father":** Paul Scanlon, "An Interview with George Lucas," *Rolling Stone,* August 25, 1977, http://www.rollingstone.com/movies/news/the-wizard-of-star-wars-20120504.

18   **"had Vader in mind":** See Taylor, *How Star Wars Conquered the Universe,* 237.

18   **"Don't tell anyone":** "George Lucas Relates to 'Lost': 'The Trick is to Pretend You've Planned the Whole Thing Out in Advance,'" *Hero Complex,* May 18, 2010, http://herocomplex.latimes.com/movies/lost-george-lucas/.

18   **"When you're creating":** Taylor, *How Star Wars Conquered the Universe,* 237.

19   **"tho I call them Mine":** H. L. Jackson, *Those Who Write for Immortality* (New Haven, CT: Yale University Press, 2015), 169, 171.

19   **here's some speculation about what might have happened:** Taylor, *How Star Wars Conquered the Universe* 248–51; also terrific is Kaminski, *The Secret History of Star Wars,* from which I have learned a great deal.

19   **"at a stroke":** Taylor, *How Star Wars Conquered the Universe,* 100.

22   **"This just seemed a really lame attempt":** Ibid., 263.

22   **"And who [Leia] ends up with":** Documentary, "The Making of Star Wars," 1977, https://www.youtube.com/watch?v=FSuDjjlIPak.

22   **"savoring the way":** George Lucas, "Star Wars: A New Hope," in *The Star Wars Trilogy* (New York: Ballantine Books, 2002), 53.

22   **"She was even more beautiful":** Ibid., 171.

22   **Here's how the novel ends:** Ibid., 260.

23   **"I want to have Luke":** Rinzler, *The Making of Star Wars,* 107.

23   **"twin sister on the other side":** Taylor, *How Star Wars Conquered the Universe,* 232.

23   **Lucas later said:** *Interviews*, 96.

23   **"enhance the audience's perception":** Taylor, *How Star Wars Conquered the Universe*, 100.

25   **Leia Brackett produced a late and mostly terrific version:** See Leigh Bracket, *STAR WARS Sequel*, http://scyfilove.com/wp-content/uploads/2010/05/Star-Wars-The-Empire-Strikes-Back-Brackett-Draft.pdf (last visited February 14, 2016).

27   **"The future is an accident":** Ryan Bradley, "Economists, Biologists, and Skrillex on How to Predict the Future," *New York Times Magazine*, November 10, 2015, http://www.nytimes.com/2015/11/15/magazine/economists-biologists-and-skrillex-on-how-to-predict-the-future.html.

27   **(He almost did):** See Phil Szostak, *The Art of Star Wars: The Force Awakens* (New York: Abrams, 2015).

27   **"I hate to say it":** See Anthony Lane, "*Star Wars: The Force Awakens* Reviewed," *New Yorker*, December 18, 2015, http://www.newyorker.com/culture/cultural-comment/star-wars-the-force-awakens-reviewed.

27   **It's a clash of two Jedi masters:** Rinzler, *The Making of Star Wars*, 64.

29   **"I was always lobbying":** Joanna Robinson, "Star Wars Writer Explains Why The Force Awakens Leaves So Many Questions Unanswered," *Vanity Fair*, December 21, 2015, http://www.vanityfair.com/hollywood/2015/12/star-wars-force-awakens-who-are-reys-parents.

29   **"It's the ship":** *Interviews*, 120.

29   **With its irresistible specificity:** Imaginative efforts have been made to restore sense. My favorite is this: Han took a shortcut. For discussion, see Amelia Hill, "Star Wars FAQ: Why Did Han Solo Say He Made the Kessel Run in 12 Parsecs?," About.com, December 10, 2014, http://scifi.about.com/od/starwarsglossaryandfaq/a/Star-Wars-Faq_Why-Did-Han-Solo-Say-He-Made-The-Kessel-Run-In-12-Parsecs.htm.

EPISODE II: THE MOVIE NO ONE LIKED:
An Expected Flop Becomes the Defining Work of Our Time

31   **broke records for nine of them:** Rinzler, *The Making of Star Wars*, 294. Over the weekend, the number of theaters showing the film expanded to 43. Taylor, *How Star Wars Conquered the Universe*, 187.

31   **four of the five:** Rinzler, *The Making of Star Wars,* 294.

31   **single-day total was $254,809:** Ibid., 295.

31   **single-day total alone was $19,358:** Taylor, *How Star Wars Conquered the Universe,* 182.

31   **Manhattan's Astor Plaza brought in $20,322:** Rinzler, *The Making of Star Wars,* 294.

31   **True, it did not quite win:** Michael Coate, "The Original First-Week Engagements of 'Star Wars,'" in10mm, http://www.in70mm.com/news/2003/star_wars/ (last visited February 13, 2016).

32   **grand weekend total of 43:** Taylor, *How Star Wars Conquered the Universe,* 187.

32   **Benton County, Oregon:** Rinzler, *The Making of Star Wars,* 304.

32   **1,100 theaters across the country:** Michael Zoldessy, "Celebrating the Original Star Wars on Its 35th Anniversary," Cinema Treasures blog, May 25, 2012, http://cinematreasures.org/blog/2012/5/25/celebrating-the-original-star-wars-on-its-35th-anniversary.

32   **for *over a year*:** Ibid.

32   **literally being worn out:** Rinzler, *The Making of Star Wars,* 304.

32   **most successful film, ever:** Ibid.

32   **from $6 a share to nearly $27:** Ibid., 302.

32   ***A New Hope* surpassed *Jaws*:** Ibid., 300. *Jaws* had earned $260 million at the box office back in 1975. *Jaws* Gross, Box Office Mojo, http://www.boxofficemojo.com/movies/?id=jaws.htm (last visited November 2, 2015).

32   **made $307 million:** Star Wars Gross, Box Office Mojo, http://www.boxofficemojo.com/movies/?page=releases&id=starwars4.htm (last visited November 2, 2015).

32   **That's 240 percent:** "Top-U.S.-Grossing Feature Films Released in 1977," IMDb, http://www.imdb.com/search/title?sort=boxoffice_gross_us&title_type=feature&year=1977,1977 (last visited January 4, 2016).

32   **It's also roughly six times:** Ibid.

32   **an estimated $1.55 billion:** "All Time Box Office," Box Office Mojo, http://www.boxofficemojo.com/alltime/adjusted.htm (last visited November 9, 2015).

32   ***Avatar*'s adjusted earnings:** Ibid.

33   **Samoa by about $700 million:** Data: Samoa, World Bank, http://data.worldbank.org/country/samoa (last visited January 4, 2016).

33   **$209 million:** *The Empire Strikes Back* Gross, Box Office Mojo, http://www.boxofficemojo.com/movies/?page=releases&id=starwars5.htm (last visited November 2, 2015).

33   **made well over $200 million:** See Star Wars Franchise Gross, Box Office Mojo, http://www.boxofficemojo.com/franchises/chart/?id=starwars.htm (last visited November 2, 2015).

33   **previous North American record:** Tre'vell Anderson and Ryan Faughnder, "'Star Wars: The Force Awakens' Now Holds Record for Largest Opening Weekend Ever," *Los Angeles Times*, December 20, 2015, http://www.latimes.com/entertainment/envelope/cotown/la-et-ct-star-wars-the-force-awakens-weekend-box-office-20151220-story.html.

33   **comparatively ordinary $85 million:** Brook Barnes, "'Star Wars: The Force Awakens' Shatters Box Office Records," *New York Times*, December 20, 2015, http://www.nytimes.com/2015/12/21/movies/star-wars-the-force-awakens-shatters-box-office-records.html.

34   ***nobody* thought":** Scanlon, "An Interview with George Lucas."

34   **"little faith in the film":** Rinzler, *The Making of Star Wars*, 36.

35   **"no applause,":** Taylor, *How Star Wars Conquered the Universe*, 156.

35   **"the board didn't have any faith":** Kirsten Acuna, "George Lucas Was Convinced 'Star Wars' Would Be a Disaster Until This Phone Call in 1977," *Business Insider*, April 18, 2015, http://www.businessinsider.com/when-george-lucas-knew-star-wars-was-a-hit-2015-4. However, Lucas had a believer in then–president of Twentieth Century Fox Alan Ladd Jr. See ibid. Lucas also had a believer in friend and fellow director Steven Spielberg. See ibid.; see also Frank Pallotta, "How Steven Spielberg Made Millions Off 'Star Wars' After a 1977 Bet with George Lucas," *Business Insider*, March 26, 2014, http://www.businessinsider.com/george-lucas-star-wars-bet-made-steven-spielberg-millions-2014-3.

35   **wasn't even worth the celluloid:** Taylor, *How Star Wars Conquered the Universe*, 184.

35   **Lucas himself projected:** Ibid., 156–57. See also Scanlon, "An Interview with George Lucas."

35   **"a zillion to one":** *Interviews*, 81.

35   **"I expected to break even":** Scanlon, "An Interview with George Lucas."

35   **His then wife:** Taylor, *How Star Wars Conquered the Universe*, 157.

36 **"had just released a flop"**: Mike Musgrove, "Review: 'How Star Wars Conquered the Universe,' by Chris Taylor," *Washington Post,* October 10, 2014, https://www.washingtonpost.com/entertainment/books/review-how-star-wars-conquered-the-universe-by-chris-taylor/2014/10/09/6cd5afa2-32bc-11e4-8f02-03c644b2d7d0_story.html.

36 **"Lucas was certain"**: Taylor, *How Star Wars Conquered the Universe,* 187.

36 **"Nobody liked it"**: Susana Polo, "Stephen Colbert and George Lucas Talk Star Wars, Wooden Dialogue, and Howard the Duck," Polygon, April 18, 2015, http://www.polygon.com/2015/4/18/8448685/stephen-colbert-george-lucas-tribeca-talk.

36 **"a complete turkey"**: *When Star Wars Ruled the World,* VH1 television broadcast, September 18, 2004, https://www.youtube.com/watch?v=1CGnXUEWFbIth.

36 **"There's this giant guy"**: Ibid.

36 **"a load of rubbish"**: Gavin Edwards, "The Many Faces of Vader," *Rolling Stone,* June 2, 2005, http://www.rollingstone.com/movies/features/the-many-faces-of-vader-20050602?page=2.

37 **"really hard to keep a straight face"**: *When Star Wars Ruled the World.*

37 **"nothing was supposed to do that"**: Carrie Fisher, "The Arrival of the Jedi," *Time,* March 31, 2003, http://content.time.com/time/specials/packages/article/0,28804,1977881_1977891_1978545,00.html.

37 **"The best I could imagine"**: Taylor, *How Star Wars Conquered the Universe,* 145.

37 **"Just wait"**: Polo, "Stephen Colbert and George Lucas Talk Star Wars."

37 **"no one predicted"**: Paul Young, "*Star Wars* (1977)," in John White and Sabine Haenni, eds., *Fifty Key American Films* (London and New York: Routledge, 2009), 177, 180.

EPISODE III: SECRETS OF SUCCESS:
Was Star Wars Awesome, Well-Timed, or Just Very Lucky?

39 **"Ultimately, we're all social beings"**: Duncan Watts, "Is Justin Timberlake a Product of Cumulative Advantage?" (2007), available at http://www.nytimes.com/2007/04/15/magazine/15wwlnidealab.t.html.

40   **"it's fun, it's delightful":** Adam Rogers, "The Force Will Be With Us. Always. — Star Wars and the Quest for the Forever Franchise," *Wired*, http://www.wired.com/2015/11/building-the-star-wars-universe/ (last visited February 13, 2016).

45   **A few years ago:** See Matthew J. Salganik et al., "Experimental Study of Inequality and Unpredictability in an Artificial Cultural Market," 311 *Science* 854, February 10, 2006, https://www.princeton.edu/~mjs3/salganik_dodds_watts06_full.pdf.

49   **Consider a mischievous experiment:** Matthew J. Salganik and Duncan J. Watts, "Leading the Herd Astray: An Experimental Study of Self-fulfilling Prophecies in an Artificial Cultural Market," 71:4 *Social Psychology Quarterly* 338, 2008, http://www.princeton.edu/~mjs3/salganik_watts08.pdf.

50   **accident, contingency, and luck:** See generally Jackson, *Those Who Write for Immortality*.

50   **"Here lies":** John Keates Tombstone, Keats-Shelley House, http://www.keats-shelley-house.org/en/writers/writers-john-keats/john-keats-tombstone (last visited February 13, 2016).

50   **In Keats's time:** Jackson, *Those Who Write for Immortality*, 149.

50   **"It seems that his reputation":** Ibid., 117.

50   **Jackson's remarkable conclusion:** Ibid., 155.

50   **"conundrum of Barry Cornwall's success":** Ibid., 161.

51   **echo chamber effects:** Ibid., 131.

51   **"easily have happened to Austen":** Ibid., 95.

51   **"were almost unknown":** Ibid., 168.

51   **"long-term survival":** Ibid., 218.

52   **"it's fun to participate":** Arion Berger, "A Night Out at the Memeplex," in Glenn Kenny, ed., *A Galaxy Not So Far Away* (New York: Henry Holt, 2002), 64.

52   **"A true mass-cultural event":** Ann Friedman, "Why Did I Pay $30 to See 'Star Wars'?," *Los Angeles Times*, December 23, 2015, http://www.latimes.com/opinion/op-ed/la-oe-1223-friedman-star-wars-mass-culture-20151223-story.html.

53   **"simultaneously a cult artifact":** Berger, "A Night Out at the Memeplex," 66.

56   **"the greatest film [he'd] ever seen":** Rinzler, *The Making of Star Wars*, 247.

56   "the greatest movie ever made": Ibid., 256.

56   In its first screening: Ibid., 288.

56   "Old people, young people, children": Taylor, *How Star Wars Conquered the Universe*, 184.

56   standstill traffic: See Rinzler, *The Making of Star Wars*, 297.

57   initial reviews were exceedingly positive: See, e.g., ibid., 296 (compiling positive reviews).

57   "most beautiful movie serial ever made": Vincent Canby, "A Trip to a Far Galaxy That's Fun and Funny," *New York Times*, May 26, 1977, http://www.nytimes.com/1977/05/26/movies/moviesspecial/26STAR.html?_r=1&.

57   "the most visually awesome": Taylor, *How Star Wars Conquered the Universe*, 164.

57   "masterpiece of entertainment": Joseph Gelmis, "Superb Sci-Fi," *Newsday*, May 27, 1977, http://www.newsday.com/entertainment/movies/star-wars-newsday-s-original-1977-movie-review-1.7922952.

57   "Every TV show news program": Taylor, *How Star Wars Conquered the Universe*, 187.

57   "[he] wanted to shoot [him]self": " 'Star Wars': Their First Time," *New York Times*, October 28, 2015, http://www.nytimes.com/interactive/2015/10/28/movies/star-wars-memories.html. See also Rinzler, *The Making of Star Wars*, 298 (compiling directors' reactions to *Star Wars*).

57   "one of the most exciting experiences": Rinzler, *The Making of Star Wars*, 298.

57   "[T]he whole world will rejoice with you": Ibid., 298.

57   Jonathan Lethem captures the feeling this way: Jonathan Lethem, "13, 1977, 21," in Kenny, ed., *A Galaxy Not So Far Away*, 1.

58   Todd Hanson does just that: Todd Hanson, "A Big Dumb Movie About Space Wizards: Struggling to Cope with *The Phantom Menace*," in Kenny, ed., *A Galaxy Not So Far Away*, 181.

59   "overwhelming[ly] popular": Gary Arnold, " 'Star Wars': A Spectacular Intergalactic Joyride," *Washington Post*, May 25, 1977, http://www.washingtonpost.com/wp-dyn/content/article/2005/04/06/AR2005040601186.html.

59   **"the year's best movie"**: See *Time*, May 30, 1977, http://content.time.com/ time/covers/0,16641,19770530,00.html.

59   **From the opening weekend**: See Rinzler, *The Making of Star Wars*, 195–96.

59   *Variety* **ran an article**: Ibid., 297.

59   **handling a hundred calls an hour**: Ibid.

59   **But he devoted time to *Star Wars***: Taylor, *How Star Wars Conquered the Universe*, 187 n. 4.

60   **"person-to-person communication"**: Rinzler, *The Making of Star Wars*, 297.

60   **"word of mouth"**: Taylor, *How Star Wars Conquered the Universe*, 189.

60   **"famous for being famous"**: Ibid.

60   **"joined an exclusive club"**: Ibid.

60   **"everything was different now"**: *Star Wars: The Legacy Revealed*, History Channel television broadcast, May 28, 2007, https://archive.org/details/ StarWarsTheLegacyRevealed2007.

60   **"It's nice to leave your niche"**: Friedman, "Why Did I Pay $30 to see 'Star Wars'?" For a more academic treatment, see Cass R. Sunstein and Edna Ullmann-Margalit, "Solidarity Goods," *Journal of Political Philosophy* 9, no. 2 (June 2001) 129.

61   **"the inevitable product"**: A. O. Scott, " 'Star Wars,' Elvis, and Me," *New York Times*, October 28, 2015, http://www.nytimes.com/2015/11/01/ movies/star-wars-elvis-and-me.html?hp&action=click&pgtype=Homepa ge&module=photo-spot-region&region=top-news&WT.nav=top-news&_ r=1&mtrref=www.nytimes.com&assetType=nyt_now.

61   **"fingerprints of the [Vietnam] war"**: Taylor, *How Star Wars Conquered the Universe*, 163. Taylor also highlights how *Star Wars* "coincided with record levels of marijuana usage among high school students; the trend would peak in 1978 and has been falling ever since." Ibid., 184.

61   **ripe for *Star Wars'* success**: David Wilkinson, *The Power of the Force* (Oxford: Lion, 2000), 67–69.

61   **"not a hopeful time in America"**: *Star Wars: The Legacy Revealed*.

61   **"evil has to be defeated"**: Ibid.

61   **"live thriftily"**: President Jimmy Carter's Report to the American People on Energy, February 2, 1977, https://www.youtube.com/ watch?v=MmlcLNA8Zhc.

EPISODE IV: THIRTEEN WAYS OF LOOKING AT STAR WARS:
Of Christianity, Oedipus, Politics, Economics, and Darth Jar Jar

71   **"Only the threat"**: George Lucas, *"Star Wars: A New Hope,"* 35.

71   **"productive of good"**: Letter from Thomas Jefferson to James Madison, January 30, 1787, http://founders.archives.gov/documents/Jefferson/01-11-02-0095.

73   **"Make no mistake"**: Jonathan V. Last, "The Case for the Empire," *Weekly Standard,* May 15, 2002, http://www.weeklystandard.com/article/2540.

73   **"It is the Empire"**: Joe Queenan, "Anakin, Get Your Gun," in Kenny, ed., *A Galaxy Not So Far Away,* 115.

74   **"Nothing in life"**: Galen Strawson, "Thinking, Fast and Slow by Daniel Kahneman – Review," *Guardian,* December 13, 2011, http://www.theguardian.com/books/2011/dec/13/thinking-fast-slow-daniel-kahneman.

74   **We display unrealistic optimism:** The best discussion is Tali Sharot, *The Optimism Bias* (New York: Pantheon Books, 2011).

74   **We dislike losses:** On loss aversion, see Eyal Zamir, *Law, Psychology, and Morality: The Role of Loss Aversion* (Oxford and New York: Oxford University Press, 2014).

76   **"In high school"**: Tina Burgess, "George Lucas' Near-Death Experience: One Moment in Heaven, a Lifetime on Earth," *Examiner,* November 7, 2012, http://www.examiner.com/article/george-lucas-near-death-experience-one-moment-heaven-a-lifetime-on-earth.

76   **"When your father clawed"**: James Kahn, *"Star Wars: Return of the Jedi,"* in *The Star Wars Trilogy,* 80.

77   **"dark journey into religious fundamentalism"**: Comfortably Smug, "The Radicalization of Luke Skywalker: A Jedi's Path to Jihad," *Decider,* December 11, 2015, http://decider.com/2015/12/11/the-radicalization-of-luke-skywalker-a-jedis-path-to-jihad/.

77   **"Within moments"**: Comfortably Smug, "The Radicalization of Luke Skywalker: A Jedi's Path to Jihad," *Decider,* December 11, 2015, http://decider.com/2015/12/11/the-radicalization-of-luke-skywalker-a-jedis-path-to-jihad/.

77   **"says a Jedi prayer"**: Comfortably Smug, "The Radicalization of Luke

Skywalker: A Jedi's Path to Jihad," *Decider,* December 11, 2015, http://decider.com/2015/12/11/the-radicalization-of-luke-skywalker-a-jedis-path-to-jihad/.

78   **"was the most erotic figure":** Lydia Millet, "Becoming Darth Vader," in Kenny, ed., *A Galaxy Not So Far Away,* 133–34, 136.

79   **"The reason Milton":** William Blake, *The Marriage of Heaven and Hell* (circa 1790).

79   **"Sooner murder an infant":** William Blake, *The Marriage of Heaven and Hell* (circa 1790).

79   **"the forces of good":** Tom Bissell, "Pale Starship, Pale Rider: The Ambiguous Appeal of Boba Fett," in Kenny, ed., *A Galaxy Not So Far Away,* 15.

80   **"Darth Vader is the fulcrum":** Queenan, "Anakin, Get Your Gun," 114.

80   **"People like villains":** Gavin Edwards, "George Lucas and the Cult of Darth Vader," *Rolling Stone,* June 2, 2005, http://www.rollingstone.com/movies/news/george-lucas-and-the-cult-of-darth-vader-20050602.

80   **"Good, I can feel your anger":** Kahn, *"Star Wars: Return of the Jedi,"* 184.

85   **"His name is unnecessary information":** Kevin O'Keeffe, "There's Either an Error in the New 'Star Wars' Crawl or a Big Surprise for Luke and Leia," *Mic,* December 24, 2015, http://mic.com/articles/131224/theres-either-an-errorin-the-new-star-wars-crawl-or-a-big-surprise-for-luke-andleia#.AzDuakWXE.

86   **Is there any wonder:** Matthew Bortolin, *The Dharma of Star Wars* (Boston: Wisdom, 2005).

88   **"The main thing":** Lance Parkin, *Magic Words: The Extraordinary Life of Alan Moore* (London: Aurum Press, 2013), 324.

89   **Consider in this light:** Michael Drosnin, *The Bible Code* (New York: Simon & Schuster, 1998).

89   **"paternicity" or "apophenia":** Michael Shermer, "Patternicity: Finding Meaningless Patterns in Meaningless Noise," *Scientific American,* December 1, 2008, http://www.scientificamerican.com/article/patternicity-finding-meaningful-patterns/.

EPISODE V: FATHERS AND SONS:
You Can Be Redeemed, Especially If Your Kid Really Likes You

91   **"He wanted me":** "In Case You've Ever Wondered About George Lucas'
          Parenting Philosophies. . .," Oh No They Didn't!, May 20, 2008, available
          at http://ohnotheydidnt.livejournal.com/23697573.html?page=4.

95   **"domineering, ultraright-wing businessman":** *Interviews,* 219.

95   **"My father wanted me:** Shawn Schaitel, "The Mythology of STAR
          WARS," YouTube, May 14, 2014, https://www.youtube.com/
          watch?v=YpiEk42_O_Q ("In this interview made in 1999 Bill Moyers
          discusses with George Lucas how Joseph Campbell and his concept of the
          Monomyth, also known as the Hero's Journey, and other concepts from
          Mythology and Religion shaped the Star Wars saga").

95   **"At eighteen":** *Interviews,* 199.

95   **"I fought him:** Ibid., 221.

96   **"You only have to accomplish":** Schaitel, "The Mythology of STAR
          WARS."

96   **"Almost all of our films":** "In Case You've Ever Wondered About George
          Lucas' Parenting Philosophies . . . ," Oh No They Didn't!, May 20, 2008,
          http://ohnotheydidnt.livejournal.com/23697573.html?page=4.

96   **"Parents try":** Ibid.

96   **"he lived to see me":** Ibid.

96   **He retired for two decades:** Ibid.

96   **"I was a great dad":** "George Lucas Reveals What He Hopes His
          Obituary Says," CBS News, December 15, 2015, http://www.cbsnews.
          com/news/star-wars-creator-george-lucas-kennedy-center-honors-
          directing-career/.

97   **In the end:** See Schaitel, "The Mythology of STAR WARS."

97   **"Only through the love":** Jim Windolf, "Star Wars: The Last Battle,"
          Vanity Fair, January 31, 2005, http://www.vanityfair.com/news/2005/02/
          star-wars-george-lucas-story.

98   **"There is some good":** Martin Luther King Jr., "Loving Your Enemies"
          Sermon, December 25, 1957, http://www.thekingcenter.org/archive/
          document/loving-your-enemies-0=#.

99   **"The boy was good":** Kahn, *"Star Wars: Return of the Jedi,"* 220.

100   **"You can type this shit":** Kyle Buchanan, *It Took Almost 40 Years, But*

Harrison Ford Is Now a Star Wars *Fan, Vulture,* July 11, 2015, http://www
.vulture.com/2015/07/after-38-years-harrison-ford-is-a-star-wars-fan.html.

100   **"I think I'm a terrible writer":** *Interviews,* 110.

100   **"I'd be the first person":** Ian Freer, "Star Wars Archive: George Lucas
1999 Interview," *Empire,* December 11, 2015, http://www.empireonline.
com/movies/features/star-wars-archive-george-lucas-1999-interview/.

100   **"George isn't the best":** *Interviews,* xii.

100   **"Let go of fear":** Matthew Stover, *Star Wars: Episode III: Revenge of the
Sith* (New York: Random House, 2005), 196.

101   **"His friends were in terrible danger":** Donald F. Glut, "*Star Wars: The
Empire Strikes Back,*" 166.

101   **"Even despair is attachment":** Stover, *Revenge of the Sith,* 363.

101   **"His undoing":** Edwards, "George Lucas and the Cult of Darth Vader."

101   **"The Stoics think":** "An Interview with Martha Nussbaum," *Philosophy
for Life,* February 5, 2009, http://www.philosophyforlife.org/an-interview-
with-martha-nussbaum/.

EPISODE VI: FREEDOM OF CHOICE:
It's Not About Destiny or Prophecy

103   **"Everybody has the choice":** Interview with Bill Moyers (1999), available
at http://billmoyers.com/content/mythology-of-star-wars-george-lucas/.

105   ***"Force Awakens, New Hope, Empire":*** Adam Rogers, "*Star Wars'*
Greatest Screenwriter Wrote All Your Other Favorite Movies Too," *Wired,*
November 18, 2015, http://www.wired.com/2015/11/lawrence-kasdan-qa/.

108   **"Luke is faced with the same issues":** Windolf, "Star Wars: The Last
Battle."

110   **"A man must follow":** Lucas, "*Star Wars: A New Hope,*" 219.

110   **"Life sends you":** Freer, "Star Wars Archive."

111   **"It is unavoidable":** Kahn, "*Star Wars: Return of the Jedi,*" 174.

111   **"You have control":** Schaitel, "The Mythology of STAR WARS."

111   **"My favorite line":** Rogers, "*Star Wars'* Greatest Screenwriter Wrote All
Your Other Favorite Movies Too."

EPISODE VII: REBELS:
Why Empires Fall, Why Resistance Fighters (and Terrorists) Rise

114 **"I started to work"**: Rinzler, *The Making of Star Wars*, 7–8, 16.

114 **"during a period when Nixon"**: "The Oppression of the Sith," *Star Wars Modern*, February 18, 2010, http://starwarsmodern.blogspot.com/2010/02/communist-manifesto-turns-160-neocons.html.

115 **"with a dozen reporters"**: "George Lucas Interview," Boston.com, http://www.boston.com/ae/movies/lucas_interview/ (transcript of Ty Burr's interview with George Lucas).

115 **"came out of conversations"**: Erin Whitney, "Kylo Ren of 'Star Wars: The Force Awakens' Was Inspired by Nazis . . . Sorta," *Huffington Post*, August 25, 2015, http://www.huffingtonpost.com/entry/kylo-ren-the-force-awakens-nazis_55dca490e4b04ae49704973c.

116 **"The Senate has surrendered"**: Stover, *Revenge of the Sith*, 152.

117 **"The Imperial Senate will no longer"**: Lucas, *"Star Wars: A New Hope,"* 42.

117 **"why . . . the senate after killing"**: David Germain, "Sci-Fi Themes Hit Closer to Home," *L.A. Times*, May 16, 2005, http://articles.latimes.com/2005/may/16/entertainment/et-starwars16.

117 **"It's the same thing"**: "Cannes Embraces Political Message in 'Star Wars,'" Associated Press, May 16, 2005, http://www.today.com/id/7873314/ns/today-today_entertainment/t/cannes-embraces-political-message-star-wars/#.Voay4JMrJp9.

117 **"The power to dissolve Parliament"**: *Hitler Empowered to Dissolve Parliament; Rule by Decree; State Ouster of Cohn Accomplished*, Jewish Telegraphic Agency, February 2, 1933, http://www.jta.org/1933/02/02/archive/hitler-empowered-to-dissolve-parliament-rule-by-decree-state-ouster-of-cohn-accomplished.

119 **"We also have to work"**: http://www.washingtonpost.com/wp-dyn/content/blog/2005/11/07/BL2005110700793.html.

119 **"I want to work with Congress"**: President Barack Obama, Weekly White House Address, May 17, 2014, https://www.whitehouse.gov/the-press-office/2014/05/17/weekly-address-working-when-congress-won-t-act.

120 **"I think there reaches a point"**: Karoun Demirjian, "Democrats Hint They Are Ready for Obama to Shut Down Gitmo Alone," *Washington Post*,

November 11, 2015, https://www.washingtonpost.com/news/powerpost/
wp/2015/11/11/democrats-hint-they-are-ready-for-obama-to-shut-down-
gitmo-alone/.

121 **"floating fancies or fashions":** Edmund Burke, *Reflections on the
Revolution in France* 95 (L. G. Mitchell ed., Oxford University Press
2009) (1790).

121 **"No one generation":** Burke, *Reflections on the Revolution in France* 95.

122 **"If we are wrong":** Martin Luther King Jr., The Montgomery Bus Boycott
Speech, December 5, 1955, http://www.blackpast.org/1955-martin-
luther-king-jr-montgomery-bus-boycott.

123 **"I know it's a long shot":** Lucas, *"Star Wars: A New Hope,"* 28.

123 **"certain twisted genius":** Lucas, *"Star Wars: A New Hope,"* 40.

123 **"Some of you still don't realize":** Lucas, *"Star Wars: A New Hope,"* 41.

123 **"Remember, Luke":** Ibid., 97.

124 **"First they came":** *Martin Niemoller: First They Came for the Socialists,*
Holocaust Encyclopedia, http://www.ushmm.org/wlc/en/article.
php?ModuleId=10007392.

125 **"predict that a spark":** Foreign Affairs Committee, House of Commons,
"British Foreign Policy and the 'Arab Spring': Second Report of Session
2012–13," Report No. HC 80, at 13 (UK). For a brief summary of the
uprisings in Tunisia, Egypt, and Libya, see ibid. at 16.

125 **The United States and Canada:** See "U.S. Intelligence Official
Acknowledges Missed Arab Spring Signs," *Los Angeles Times*, World Now
blog, July 19, 2012, http://latimesblogs.latimes.com/world_now/2012/07/
us-intelligence-official- acknowledges-missed-signs-ahead-of-arab-
spring-.html; Stephanie Levitz, "Arab Spring Caught Canada by
Surprise: Government Report," *Huffington Post*, May 6, 2013, http://
www.huffingtonpost.ca/2013/05/06/arab-spring-canada-government-
report_n_3224719.html.

125 **"vast majority of academic specialists":** F. Gregory Gause III, "Why Middle
East Studies Missed the Arab Spring: The Myth of Authoritarian Stability,"
Foreign Affairs, July/Aug. 2011, https://www.foreignaffairs.com/articles/north-
africa/2011-07-01/why-middle-east-studies-missed-arab-spring.

125 **"We know that something":** Jeff Goodwin, "Why We Were Surprised (Again)
by the Arab Spring," *Swiss Political Science Review* 17 (2011): 452, 453.

127   **"Biggs is right":** Lucas, *"Star Wars: A New Hope,"* 54.

128   **with just three steps:** Susanne Lohmann, "The Dynamics of Informational Cascades: The Monday Demonstrations in Leipzig, East Germany, 1989–91," *World Politics* 47 (October 1994): 42.

128   **"Luke, I'm not going":** Lucas, *"Star Wars: A New Hope,"* 34.

131   **not knowing what people really think:** For a terrific discussion, on which I have drawn here, see Timur Kuran, *Private Truths, Public Lies* (Cambridge, MA: Harvard University Press, 1998).

132   **Group polarization:** See Cass R. Sunstein, *Going to Extremes* (Oxford and New York: Oxford University Press, 2009).

136   **"in lowly stations":** Gordon S. Wood, *The Radicalism of the American Revolution* (New York: Knopf, 1991), 29–30.

137   **"they've got a photography school":** Polo, "Stephen Colbert and George Lucas Talk Star Wars."

137   **"I got there on a fluke":** *Interviews*, 65.

137   **"What! You mean you can":** Polo, "Stephen Colbert and George Lucas Talk Star Wars."

138   **"one obscure Tunisian's actions":** Philip E. Tetlock and Dan Gardner, *Superforecasting: The Art and Science of Prediction* (New York: Crown, 2015), 10.

139   **"little space thing":** *Interviews*, 92.

139   **"Harrison had done":** Taylor, *How Star Wars Conquered the Universe*, 146.

EPISODE VIII: CONSTITUTIONAL EPISODES:
Free Speech, Sex Equality, and Same-Sex Marriage as Episodes

148   **Even given the backdrop:** "We Ask 10 Sci-Fi Authors to Write Star Wars: Episode VII," *Popular Mechanics*, May 21, 2014, http://www.popularmechanics.com/culture/movies/g1523/we-ask-10-sci-fi-authors-to-write-star-wars-episode-vii/.

151   **"living Constitution":** See, e.g., David A. Strauss, "Do We Have a Living Constitution?," *Drake Law Review* 59 (2011): 973.

152   **arresting metaphor of a chain novel:** See Ronald Dworkin, *Law's Empire* (Cambridge, MA: Belknap Press, 1985), 229–39.

154   **"We begin with several propositions":** *Virginia State Board of Pharmacy v. Virginia Citizens Consumers Council,* 425 U.S. 748, 761, 770 (1976).

155   **As late as 1963:** See, e.g., *Dennis v. United States,* 341 U.S. 494 (1951), which remained good law until *Brandenburg v. Ohio,* 395 U.S. 444 (1969).

155   **most prominently in 1964:** *New York Times Co. v. Sullivan,* 376 U.S. 254 (1964).

155   **and 1969:** *Brandenburg,* 395 U.S. 444.

155   **"the pall of fear and timidity":** *Sullivan,* 376 U.S. at 278.

155   **Star Trek does:** "Mirror Universe," Wikia, http://en.memory-alpha.org/wiki/Mirror_universe (last visited January 4, 2016).

155   **that principle is a product of the 1950s:** See *Brown v. Board of Education* (Brown I), 347 U.S. 483 (1954).

156   **is a product of the 1960s:** See *Engel v. Vitale,* 370 U.S. 421 (1962).

156   **comes from the 1970s:** See *Califano v. Goldfarb,* 430 U.S. 199 (1977).

156   **are a product of the 1990s and 2000s:** See *Grutter v. Bollinger,* 539 U.S. 306 (2003).

156   **until the twenty-first century:** See *District of Columbia v. Heller,* 554 U.S. 570 (2008).

156   **Originalism:** There are many different varieties. I am speaking of the kind defended in Antonin Scalia, *A Matter of Interpretation* (Princeton, NJ: Princeton University Press, 1998), rather than in Jack Balkin, *Living Originalism* (Cambridge, MA: Belknap Press of Harvard University Press, 2012). In fact, Balkin's account fits well with what I am suggesting here.

157   **purposes of the provisions:** Stephen Breyer, *Active Liberty* (New York: Knopf, 2005), can be read in this way, though I think his argument is subtler.

EPISODE IX: THE FORCE AND THE MONOMYTH:
Of Magic, God, and Humanity's Very Favorite Tale

161   **"Throughout the inhabited":** Joseph Campbell, *The Hero with a Thousand Faces* (New World Library, 2008), 1.

163   **George Akerlof and Robert Shiller . . . *Phishing for Phools*:** George Akerlof and Robert Shiller, *Phishing for Phools: The Economics of Manipulation and Deception* (New Jersey: Princeton University Press, 2015).

163   **Daniel Kahneman . . . *Thinking, Fast and Slow***: Daniel Kahneman, *Thinking, Fast and Slow* (New York: Farrar, Straus and Giroux, 2011).

164   **Robert Cialdini's *Influence***: Robert Cialdini, *Influence: The Psychology of Persuasion* (New York: William Morrow, 1984).

166   **"In my experience"**: Lucas, "*Star Wars: A New Hope*," 147.

166   **"superforecasters"**: See Tetlock and Gardner, *Superforecasting.*

168   **report a frequent conversation**: Christopher Chabris and Daniel Simons, *The Invisible Gorilla: How Our Intuitions Deceive Us* (New York: Crown, 2009), 6.

169   **"We experience far less"**: Ibid., 7.

169   **"We have only so many"**: Mariette DiChristina, "How Neuroscientists and Magicians Are Conjuring Brain Insights," *Scientific American*, May 14, 2012, http://blogs.scientificamerican.com/observations/how-neuroscientists-and-magicians-are-conjuring-brain-insights/.

170   **"Attention is like water"**: Adam Green, "A Pickpocket's Tale," *The New Yorker*, January 7, 2013, http://www.newyorker.com/magazine/2013/01/07/a-pickpockets-tale.

170   **David Jaher's *The Witch of Lime Street***: David Jaher, *The Witch of Lime Street: Séance, Seduction and Houdini in the Spirit World* (New York: Crown Publishers, 2015). The various quotations in the following pages are from Jaher's captivating book.

174   **"leap of faith"**: Schaitel, "The Mythology of STAR WARS."

175   **"If there's only one God"**: Ibid.

175   **"I wanted a concept of religion"**: Ryder Windham, *Star Wars Episode I: The Phantom Menace Movie Scrapbook* 11 (New York: Random House, 1999).

175   **"is telling an old myth"**: Schaitel, "The Mythology of STAR WARS."

175   **"last mentor"**: Ibid.

176   **"When I did Star Wars"**: Ibid.

176   **"You know, I thought real art stopped"**: Taylor, *How Star Wars Conquered the Universe*, 278.

176   **"With Star Wars, it was the religion"**: Trent Moore, "George Lucas Tries to Explain the Real Meaning of the Star Wars Saga," *Blastr*, October 27, 2014, http://www.blastr.com/2014-10-27/george-lucas-tries-explain-real-meaning-star-wars-saga.

EPISODE X: OUR MYTH, OURSELVES:
Why Star Wars Gets to Us

177   **There's a deep human desire:** See Michael Chwe, *Rational Ritual: Culture, Coordination, and Common Knowledge* (Princeton, NJ: Princeton University Press, 2013).

179   **"feel the rapture":** Ep. 2: Joseph Campbell and the Power of Myth— "The Message of the Myth," Moyers & Company, March 8, 2013, http://billmoyers.com/content/ep-2-joseph-campbell-and-the-power-of-myth-the-message-and-the-myth-audio/.

179   **"is twice blest":** William Shakespeare, *The Merchant of Venice*, act 4, sc. 1.

INDEX

A

Abrams, J. J.
  on adventure, 6
  amazing and spectacular, 142
  brand and, 41
  choices in *The Force Awakens*, 27
  devil's party and, 178
  Facebook speculations, 132
  Jeffersonian theme, 71
  as judge and creator, 147
  on own path, 148
  what if questions, 115
Adele, 80
adventure, 6, 15, 17, 42, 57, 61, 175
agency, 18–19, 179
Akerlof, George, 163
Algeria, 125
American Graffiti, 35, 139
American Revolution, 131, 132
Anakin Skywalker (character)
  attachment, 100–102
  attachment and, 96–97
  choices, 106
  as clueless, 70
  decision in *Revenge of the Sith*, 108–109
  as father, 97–100
  feeling anger, 80
  final battle, 144
  killing Satan, 67
  love scenes, 26
  as monster, 176
  as part of Holy Trinity, 67
  path to Dark Side, 68, 76–77,
      109-110, 144
  redemption of, 96–97, 101–102
  return to Light Side, 97
  virgin birth product, 66
apartheid, 44, 131
Apocalypse Now, 114
apophenia, 89
Arab Spring
  heroes, 120
  human rights and, 7
  injustice and, 131
  as surprise, 124–126

attachment
    A. Skywalker, 96–97, 100–102
    L. Skywalker, 101
    nature of, 5–6
Attack of the Clones
    crawl of, 139–140
    power in, 118
    ranking, 143
    visions in, 108
attention, 169–170
Austen, Jane, 9, 49–51
availability heuristic, 131
Avatar, 32–34
Avco Theater, 56
Awake, 45

B

Bacon, Francis, 89
bandwagon effect, 40–41
BB-8 (character), 76, 179
Beatles, 2, 40, 43, 80
behavioral science, 7, 10, 73–76
beliefs, 126, 130, 167
Berger, Arion, 52–53
betrayal, 121, 149
Biggs (character), 123, 128–129, 131
Bird, James Malcolm, 171
Blake, William, 19, 49, 51, 79–80, 178
blindness, 126–128, 169
Bonaparte, Napoleon, 115
Bortolin, Matthew, 86
Boston Herald, 171
Brackett, Leigh, 25–27
Bradbury, Ray, 136

Brady, Tom, 164–165
brain bandwidth, 169
brand, 41
A Bridge Too Far, 32
Brunton, Mary, 49–51
Buddhism, 85–86, 101, 178
Burke, Edmund, 121–122
Burtt, Ben, 37
Bush, George W., 118–119
butterfly effect, 136–139
Byatt, A. S., 20

C

Calefaction, 46
Campbell, Joseph
    on art, 176
    on gods, heavens, hells, 1
    as Lucas mentor, 5, 100, 175, 179
    monomyth of, 66, 179
    on myth, 161
Canby, Vincent, 57
capitalism, 4
Carter, Jimmy, 61
causal chains, 137
Chabris, Christopher, 168–169
chain novel, 152–153
chaos, 72–73
Cheney, Dick, 119
Chewbacca (character), 23, 31, 36
Christianity, 4, 66–67, 178
Cialdini, Robert, 164
civil rights movement, 61–62, 103–104,
    134–135
Civil War, 155, 157

Cleg (character), 14
climate change, 119
Clinton, Hillary, 2, 34
Close Encounters of the Third Kind, 32
Coben, Harlan, 20
Colbert, Stephen, 60
Cold War, 42, 61, 105
communism
    fall of, 7
    prevailing, 105
    rise of, 89
Congress, 118–120
conspiracy theories, 84, 87–89, 133
Constitution, U.S., 146–147, 149–151,
    155
constitutional law
    aura of inevitability, 157
    episode order, 157–159
    following rules and, 149–152
    "I am your father" moment and, 146
    infinite possibilities, 148–149
    Jedi Knights and, 157
    legal style, 153–156
    originalism and, 156–157
    overview, 145–147
    Sith and, 157
    as Star Wars episodes, 152–153
Cornwall, Barry, 49–51
Coronet Theater, 36, 56
Crabbe, George, 49–51
Crandon, Mina, 171–173
The Creation of the American Republic
    (Wood), 146
Cronkite, Walter, 59
cruel and unusual punishment, 151
Cruz, Ted, 2

C-3PO (character), 16
The Cuckoo's Calling (Galbraith), 48–49
cultural goods, 52
cultural resonance
    A New Hope and, 45
    as success secret, 41, 44–45
cultural swoon, 52

D

da Vinci, Leonardo, 9, 59
Daniels, Anthony, 36
Dark Side. *See also* Sith
    A. Skywalker's, 68, 76–77, 109–110,
        144
    avoiding, 173
    Cheney and, 119
    devil's party and, 78–84
    Light Side compared to, 65, 105
    owning your own shadow, 81
    seduction by, 21
    as sexy, 79
    in Star Trek, 80–82
    turning from, 26
    visiting, 5
    Yoda despair of, 101
Darth Sidious (character), 109. *See also*
        Palpatine (character)
Darth Vader (character), 3, 10. *See also*
        Anakin Skywalker (character)
    death of, 99
    as erotic, 79
    failing as Jedi Knight, 101
    as father, 94–96
    fierce and terrifying, 94

Darth Vader *(continued)*
    as general, 16
    Lucas on, 11
    as most memorable character, 78–80
    optimistic bias, 75
    part human, part machine, 76
    on ruling galaxy, 19
    tragedy of, 12, 66
    turning of, 124
    working on mind, 174
de Vere, Edward, 89
Death Star, 84, 86, 165
dehumanizing, 76
democratic systems, 116–117, 120
destiny
    controlling, 85
    freedom of choice and, 110–111
    fulfilling, 98, 109
    honoring, 180
    Lucas on, 111
    in *A New Hope*, 47
devil's party
    Abrams and, 178
    Dark Side and, 78–84
    Lucas and, 178
The Dharma of Star Wars (Bortolin), 86
Dickens, Charles, 40–41, 47
dictatorships, 117
Dodds, Peter, 45–48
Dooku (character), 139–140
down look, 135–136
Doyle, Arthur Conan, 170–172
droids, 3, 35, 42, 76, 100, 163
Dworkin, Ronald, 152–153
Dylan, Bob, 63–64

E

echo chamber effect, 51
economic policy, 119
effervescent giddiness, 30
Egypt, 125
Ellerbee, Linda, 61
Empire
    fall of, 128
    Nixon and, 41–42
    opposing, 113
    Soviet Union as, 42
    turbulence and, 71
    tyrannical, 134
    U.S. as, 41
The Empire Strikes Back, 6
    as best, 142
    early script, 25–27
    evil triumphing, 79–80
    financial success, 33
    "I am your father" moment, 19, 21
    kiss in, 23
    ranking, 143
    sexual charge in, 24
    writing of, 22
equal protection, 151, 155–156, 157
E.T., 33

F

Facebook, 52, 132
fanaticism, 123
fascist systems, 116
fast thinking (System 1), 163–164, 166

father(s)
  A. Skywalker, 97–100
  Darth Vader as, 94–96
  as Jedi Knights, 94
  Lucas as, 96
  Lucas on own, 95–96
  obsession with, 5
  overview, 91–93
  regret and, 93–94
  as Sith, 94
  Solo as, 97
  Spielberg and, 96
  tale of, 13
feelings, 165–166
feminism, 69–71
Finn (character)
  choices, 106
  as irresistible, 177
First Amendment, 154–155
First Order, 71, 74, 113, 122, 150
Fisher, Carrie, 37
Flash Gordon (character), 11–13, 18,
      177, 180
Flynn, Gillian, 20, 44
Force. *See also* Light Side
  Force Ghosts, 170
  as God, 4
  knowledge of, 3
  mastering, 161–162
  may the Force be with you, 2
  two sides of, 5
Force, monomyth and
  invisible gorilla experiment and,
      168–170
  Jedi nudges, Sith nudges, 173–174

  old myth, new way, 174–176
  overview, 161–164
  pattern recognition, 164–166
  superforecasters and, 166–167
  supernatural abilities and, 167
  twentieth century Jedi, 170–173
The Force Awakens
  Abrams on, 115
  appearance of, 2
  brand and, 41
  choices in, 27
  collaboration on, 6
  disappointment in, 132–133
  droids, 76
  financial success, 33–34
  General Leia, 70
  informational cascades, 53
  as mass-cultural event, 52–53
  as national celebration, 178
  network effects, 63
  Oedipal theme, 69
  opening and success, 142
  ranking, 143
  sex equality and, 69
  Solo as father, 97
  Solo's death in, 29
  success of, 1, 34
  timing, 62
  triumph or disappointment, 7
  as unifying, 3
Ford, Harrison, 29–31, 36, 78, 138–139
Foster, Alan Dean, 22
framing, 163
Frankfurt, Harry, 107–109
free will, 107–108, 174

freedom of choice
    destiny and, 110–111
    emphasis on, 6
    human condition, 180
    insistence on, 179
    liberty and autopilot, 105–106
    overview, 103–105
    philosophy, 107–108
    tale of, 17
freedom of speech, 150–152, 154
French Revolution, 115, 120, 131
Freud, Sigmund, 67, 179
Friedman, Ann, 52, 60
Frost, David, 61
Frost, Robert, 9

G

Galbraith, Robert, 48–49, 53–54, 63
Gardner, Dan, 166
Gelmis, Joseph, 57
Gingrich, Newt, 61
Gone Girl (Flynn), 20, 44
Gone with the Wind, 33
Goodwin, Jeff, 125
Google, 2
Great Recession of 2008, 62, 177
Greedo (character), 85
group polarization, 131–135
Guinness, Alec, 35, 37
gun control, 119, 156

H

Hamill, Mark, 22, 23–24, 36–37
Hamilton, 146
Han Solo (character), 10, 14, 23
    choices, 106
    as clueless, 70
    death of, 29
    as father, 97
    as irresistible, 177
    Leia on, 138
    optimistic bias, 75
    as Padawan, 84–85
    as rogue, 78
    status quo and, 129
Hanson, Todd, 58
Harry Potter, 42, 44, 45–46, 176
heroism, 5, 100, 103, 120, 175–176
Hero's Journey, 6, 176
The Hero with a Thousand Faces
    (Campbell), 5, 100
The Hidden Fortress, 15–16
Hitler, Adolf, 117–119, 124, 137
The Hobbit: An Unexpected Journey, 33
Holocaust, 89
Houdini, Harry, 171–172
How Star Wars Conquered the Universe
    (Taylor), 19
human attachment, 6
The Hunger Games, 42, 44, 176
Hunt, Leigh, 49–51
Hydraulic Sandwich, 46

I

"I am your father" moment, 19–22, 146, 158
immigration reform, 119
Imperial Star Destroyer, 64
inattentional blindness, 169
incommensurability, 83
Influence (Cialdini), 164
information pool, 134
informational cascades
   The Force Awakens, 53
   L. Skywalker and, 53–54
   A New Hope, 53, 55
   Obama and, 140
   success secrets, 53–55
intuition, 137, 165, 167
investigative magic, 7
invisible gorilla experiment, 168–170, 173
Islam, 134

J

Jackson, H. J., 49–51
Jackson, Peter, 57
Jaher, David, 170
James, LeBron, 80
Jar Jar Binks (character), 87, 134
Jaws, 32, 59
Jedi Knights, 18, 59
   aspiring to be, 88
   clash of, 27–28
   constitutional law and, 157

Darth Vader failing as, 101
fast thinking (System 1) and, 166
fathers as, 94
feelings and, 165–166
group polarization and, 132
in Jihad, 77
mind tricks, 150, 163, 169, 173
nudges, 173–174
Oedipus Jedi, 67–69
order and, 72–73
pattern recognition and, 165
physical manifestations, 171
real-world code, 166–167
role of, 157
slow thinking (System 2) and, 164
twentieth century Jedi, 170–173
Jefferson, Thomas, 71
Jihad, 77
Jobs, Steve, 2, 9
Johnson, Magic, 165
Jordan (nation), 125
Jordan, Michael, 80
Journal of the Whills, 149, 151, 156
Joyce, James, 167
Jung, Carl, 81
Jurassic World, 33

K

Kahneman, Daniel, 10, 74, 163
Kaminski, Michael, 17
Kane (character), 14
Kant, Immanuel, 80

Kasdan, Lawrence
  argument with, 28–29
  brilliance, 27
  character debate, 153
  collaborator, 6
  favorite lines, 111
  on fulfilling what's inside you,
    105–106
  on *A New Hope*, 40
  on own path, 148
Keats, John, 49–51
Kennedy, Anthony, 145
Kennedy, John F., 41, 84, 113
Kennedy, Robert, 41
Kessel Run, 29–30, 59
King, Martin Luther, Jr.
  assassination, 41
  as conservative, 6–7
  on good and evil, 98
  let freedom ring speech, 103–104
  on Montgomery Bus Boycott, 122
  as rebel, 120–122
King Lear (Shakespeare), 40, 89
Kingdom of the Spiders, 32
Kubrick, Stanley, 65
Kuran, Timur, 130
Kurosawa, Akira, 15–16
Kurtz, Gary, 13
Kylo Ren (character), 52, 78, 85, 99,
  106, 178

L

Lando Calrissian (character), 24
Lane, Anthony, 27

Lars (character), 14
Leia Organa (character)
  favorable cascades, 141
  as General Leia, 69–70
  on Han, 138
  as rebel, 121, 128
  sense of grievance, 131
  as twin, 22–24, 85
Lethem, Jonathan, 57–58
LGBT rights, 135
liberty, 150
Libya, 125
Light Side. *See also* Force
  A. Skywalker returning to, 97
  choosing, 5
  compared to Dark Side, 65, 105
  in favor of, 4
  in Star Trek, 80–82
lightsabers, 3, 10, 35, 69, 75, 179
Lincoln, Abraham, 2, 80
Lippincott, Charley, 36, 48, 56
Lohmann, Susanne, 128
"Long Time Comin'," 106
long-term survival, 51
Lorenz, Edward, 138
Lucas, George
  attending film school, 137–138
  bandwagon effect and, 40–41
  Campbell as mentor, 5, 175,
    179
  character debate, 153
  character names, 13–15
  on characters taking over, 18–19
  choice-making, 27–29
  collaborators, 6
  conveying spiritual, 175

creating twins, 22–24
on Darth Vader, 11
on destiny, 111
devil's party and, 178
on dictatorships, 117
early drafting, 15–16
effervescent giddiness, 30
episode order and, 158
as father, 96
favorable cascades, 141
fearing catastrophe, 4
finest narrative, 102
Flash Gordon and, 11–13
on heroism, 103
hiring Ford, 139
human situations and, 100
idea for saga, 7
intentions for young people, 41
as judge and creator, 147
on keeping eyes open, 110
kernels of thinking, 24
as L. Skywalker, 79
on myth, 176
on Nixon, 114–115
overview of journey, 9–10
on own father, 95–96
parsecs and, 29–30
plan for Binks, 87
plot and, 156
point of view, 21–22
political ideas, 113–114
redemption and, 17
retirement, 96
on *Revenge of the Sith*, 115
running out of money, 35
shivers down the spine, 19–21
on Star Wars conception, 11–12
technology and, 76
on vacation, 36
vision, 71
visuals and, 21
writing of, 12
luck
    A New Hope and, 48
    Obi-Wan on, 166
    timing and, 177
Luke Skywalker (character), 10
    attachment and, 101
    choices, 106
    as Christ figure, 67
    as clueless, 70
    concern for, 19
    first appearance, 16
    gift of, 96–97
    inattentional blindness, 169
    informational cascades,
        53–54
    Lucas as, 79
    oath by, 26
    Obama as, 113
    Obi-Wan lying to, 21
    as Oedipus, 68–69
    as part of Holy Trinity, 67
    radicalization, 77
    rage of, 98
    as rebel, 120–121, 127–128,
        135
    resistance by, 124
    as son, 97–100
    status quo and, 75, 129
    as twin, 22–24, 85
Lumpawaroo (character), 87

M

Mace Windu (character), 14, 109
Mad Max, 44
Mad Men, 44
Madison, James, 154
Malcolm X, 41
Mandela, Nelson, 98
Mann, Thomas, 167
Mann's Chinese Theatre, 31
Marlowe, Christopher, 89
mass-cultural event, 52–53
The Matrix, 176
McComas, Henry C., 172
metaphorical killing, 21
Michelangelo, 40
Mill, John Stuart, 80
Millennium Falcon, 3, 10, 29–30
Millet, Lydia, 78–80
Milton, John, 79–80
Mona Lisa, 63
monomyth, 66, 175, 180. *See also* Force,
    monomyth and
Monroe, 14
Montgomery Bus Boycott, 122
Moore, Alan, 87–88
Moore, Julianne, 80
Mozart, Wolfgang Amadeus, 40–41, 47
Munn, Orson, 170–171
Music Lab, success secrets and, 45–48,
    51–55
Muslims, 134
myth. *See also* Force, monomyth and;
    monomyth
    Campbell on, 161

of creative foresight, 10
Lucas on, 176
modern, 4–5, 176
our myth, ourselves, 177–180
mythology, 60

N

national polls, 141
NATO, 138
network effects
    The Force Awakens, 63
    A New Hope, 59
    success secrets, 52–53
New York, New York, 36
New York Times, 57, 171
A New Hope
    as big lift, 177
    cultural resonance and, 45
    Death Star plans, 84
    defining our time, 31–37
    destiny, 47
    droids, 76
    early apathy toward, 34–35
    early drafts, 17
    first release, 3–4
    good fortune, 63
    greatest film ever seen, 56–58
    Imperial Senate, 117
    informational cascades, 53, 55
    Kasdan on, 40
    luck and, 48
    marketing enterprise, 36
    mysteries in, 158

network effects, 59
novelization, 22, 123
Obi-Wan lying to L. Skywalker, 21
opening scene, 64
Oscar nominations, 57
plot of, 13, 156
popularity and financial success,
    32–33
quality and, 56
ranking, 143
release of, 18, 22
reputational cascades, 59
reviews, 57–58
sexual charge in, 24
social influences and, 59–60
Spielberg and, 40
success of, 7
timing, 60–62
writing of, 12
A New Hope (Foster), 22, 123
Newsday, 57
The New Revelation (Doyle), 170
Niemöller, Martin, 124
Nixon, Richard
    Empire and, 41–42
    Lucas on, 114–115
    Palpatine and, 114
    resignation, 104
nondelegation doctrine, 118
Nudge (Thaler and Sunstein), 174
nudges, 173–174

Obama, Barack
    election, 132
    endgame, 84
    executive power, 119–120
    informational and reputational
        cascades, 140
    as L. Skywalker, 113
    parents of, 137
    political campaign, 140–141
    Putin and, 3
    Reagan and, 80
    on Star Wars, 2
Obi-Wan Kenobi (character)
    concern for L. Skywalker, 19
    on deception, 163
    final battle, 144
    gentle and calming, 94
    inattentional blindness, 169
    knowing smile, 85
    on luck, 166
    lying to L. Skywalker, 21
    praying to, 3
    radical cause, 77
    on rebels, 123
    smiling before struck down, 83
    teaching, 86
    as wizened, 175
    working on mind, 174
Oedipus Jedi, 67–69
opinion polls, 130
optimistic bias, 75
order, 72–73

originalism, 156–157
Oswald, Lee Harvey, 88
The Other Side of Midnight, 36

P

Padmé Amidala (character)
on democracy, 120
on loss of freedom, 115
love scenes, 26
as part of Holy Trinity, 67
solid as rock, 70
Palpatine (character)
blindness of, 126
as clueless, 124
dictator, 73
feeling anger, 80
as Hitler, 117–119
Nixon and, 114
optimistic bias, 75
reference to, 52
as revolutionary, 122
secret hero, 72
seizure of authority, 117
Sith Lord, 17, 21–22
triumph over, 109
working on mind, 174
Paradise Lost (Milton), 79
parsecs, 29–30
pattern recognition, 164–166
patternicity, 89
The Phantom Menace
arc of plot, 158–159
financial success, 33
ranking, 143

phishermen, 163, 174
Phishing for Phools (Akerlof and Shiller), 163
phools, 163
Picasso, Pablo, 167
planning fallacy, 10
Plummer, John, 137–138
pluralistic ignorance, 131
Poe Dameron (character), 152, 177
point of view, 21–22
politics
culture and, 1
Lucas political ideas, 113–114
political campaigns, 139–141
political speech, 154–155
polygamous marriage, 153
Possession (Byatt), 20
pragmatism, 120
"Predictability: Does the Flap of a Butterfly's Wing Set Off a Tornado in Texas?" (Lorenz), 138
premeditation, 19
Prince of Bebers (character), 13–14
Private Truths, Public Lies (Kuran), 130
Prowse, David, 36
psychic phenomena, 171–172
psychology, 163–164
Putin, Vladimir, 3, 72, 84

Q

quality
A New Hope and, 56
as success secret, 40

Qui-Gon (character), 68, 86, 174

R

R2-D2 (character), 16, 20, 76
racial segregation, 155, 157
randomness, 167
Reagan, Ronald, 1–2, 80, 105, 113,
    132, 137
reality, 83–84
rebellion
    cascades of, 128–129
    lessons, 4–5, 7
    need for, 71–72
    reputational cascades and,
        135–136
    Sith and, 126, 129
    spurring of, 132
    success of, 131
rebels
    blindness and, 126–128
    butterfly effect and, 136–139
    cascades of rebellion, 128–129
    concentration of power and,
        116–118
    conservative, 120–122
    delegating power, 118–120
    down look, 135–136
    fading idealism, 120
    group polarization and, 131–133
    heart of Star Wars, 189
    King as, 120–122
    L. Skywalker as, 120–121,
        127–128, 135
    Leia as, 121, 128

    Obi-Wan on, 123
    overview, 113–115
    political campaigns and, 139–141
    ranking of movies, 142–144
    reasons for polarization, 133–135
    Rebel Alliance, 123
    seizure of authority, 117
    status quo and, 129
    thinking of, 129–131
    unanticipated revolutions,
        123–126
reciprocity, 164
redemption
    of A. Skywalker, 96–97, 101–102
    Lucas and, 17
    in *Return of the Jedi*, 144
    theme, 6–7
regret, 93–94
religion, 175–176
religious liberty, 156
republics, 5, 7, 71
reputation, success secrets and, 49–51
reputational cascades
    A New Hope, 59
    Obama and, 140
    rebels and, 135–136
    success secrets, 55
Return of the Jedi, 6
    bikini scene, 69
    cultural impact, 43–44
    early drafts, 17
    financial success, 33
    as ingenious, 143
    ranking, 143
    redemption in, 144
    writing of, 24, 27

Revenge of the Sith
  A. Skywalker's decision in, 108–109
  Emperor in, 44–45
  Lucas on, 115
  massive drama, 158
  ranking, 143
Rey (character)
  choices, 106
  as complicated, 70–71
  as fabulous, irresistible, 142, 177
  as hero, 69
  inattentional blindness, 169
  motivations, 178
  status quo bias, 75
Rinzler, J. W., 60
ritual, 122
Robbins, Apollo, 169–170
Roddenberry, Gene, 80
Rodriguez, Sixto, 42–44, 47, 50–51, 63, 127
Roland (character), 14
Rolling Stones, 43, 80
romance, 42
Roosevelt, Franklin Delano, 80
Rowling, J. K., 9, 45, 48

S

Salganik, Matthew, 45–48, 49
same-sex marriage, 133
San Francisco Chronicle, 57
Saudi Arabia, 125
Scalia, Antonin, 147, 150–151
scarcity, 164
science fiction, 42, 60

Scientific American, 170–171
Scorsese, Martin, 36
Scott, A. O., 61
Scott, Ridley, 57
Searching for Sugar Man, 42–44
"Separation Anxiety," 46
separation of powers, 146
sex discrimination, 146, 156, 157
Shakespeare, William, 2, 9, 40–41, 47, 59, 88–89
Shiller, Robert, 163
shivers down the spine, 19–21
Simons, Daniel, 168–169
Sinatra, Frank, 40–41
Sith. *See also* Dark Side
  choosing, 106
  constitutional law and, 157
  fathers as, 94
  group polarization and, 132
  lords, 17, 79, 87, 109–110
  mind tricks, 169
  moral relativists, 22
  nations run by, 121
  nudges, 173–174
  pattern recognition and, 165
  phishermen and, 163, 174
  rebellion and, 126, 129
  revenge, 101
  untruthfulness and, 21
slow thinking (System 2), 163–164
Smokey and the Bandit, 31–32
social dynamics, 43, 45, 126, 129, 131
social influences
  group polarization and, 134
  A New Hope and, 59–60
  as success secret, 40–41

social media, 2
social proof, 164
sons
    L. Skywalker, 97–100
    obsessions of, 5
    overview, 91–93
    regret and, 93–94
    Spielberg and, 96
The Sound of Music, 33
Southey, Robert, 49–50
Soviet Union
    break up, 124
    as Empire, 42
    international affairs and, 61
    shadow of, 103
    struggle with, 105
Spielberg, Steven
    fathers, sons and, 96
    A New Hope and, 40
Spiritualism, 170
Springsteen, Bruce, 106
Star Cars, 49, 55
Star Trek
    Dark and Light Sides in, 80–82
    Star Wars and, 7, 82–83, 134
Star Wars
    Buddhist, 85–86
    constitutional law as episodes,
        152–153
    deepest lessons, 6
    enduring triumph, 176
    focus determines reality, 83–84
    franchise, 1
    imagined episodes, 148–149
    inviting speculation, 65
    Lucas on conception, 11–12

    as modern myth, 4–5, 176
    Obama on, 2
    rebel heart, 189
    religion and, 175–176
    as ritual, 122
    social media and, 2
    space opera, 179
    Star Trek and, 7, 82–83, 134
    as timeless, 7
    worldwide phenomenon, 2
Star Wars Infinities, 148
Star Wars: The Legacy Revealed, 61
status quo, 75, 129
Stevens, Wallace, 66
Stoicism, 101–102
Streep, Meryl, 80
subliminal advertising, 174
success secrets
    The Cuckoo's Calling and, 48–49
    cultural resonance, 41, 44–45
    informational cascades, 53–55
    Music Lab and, 45–48, 51–55
    network effects, 52–53
    quality, 40
    reputation and, 49–51
    reputational cascades, 55
    Searching for Sugar Man and,
        42–44
    social influences, 40–41
    summary, 63–64
    timing, 41–42
Sugar Man, 42–44, 127
superforecasters, 166–167
supernatural abilities, 167
Supreme Court, 146–147, 151, 154–157
Swift, Taylor, 2, 40–41, 44, 80

T

Taylor, Chris, 19, 60
technology, 76
The Ten Commandments, 33
terrorism, 133, 177
    September 11, 2001 terrorist
        attacks, 42, 44, 84, 88,
        89
    war on terror, 118–119
Tetlock Philip, 138, 166
Thaler, Richard, 173–174
Thinking, Fast and Slow (Kahneman),
    74, 163
Thorpe (character), 13–14
Time magazine, 59
timing
    as everything, 6
    The Force Awakens, 62
    luck and, 177
    A New Hope, 60–62
    as success secret, 41–42
Titanic, 33–34
"Trapped in an Orange Peel," 46
Trump, Donald, 72
Tunisia, 125, 138
Tversky, Amos, 10
Twentieth Century Fox, 32, 34–35,
    56
Twitter, 2
2001: A Space Odyssey, 57, 65

V

Valorum (character), 14
Variety, 57, 59
Vietnam War, 61, 114
von Hindenburg, Paul, 118

W

Walker, Scott, 140–141
Washington, George, 9, 72
Washington Post, 59
Watchmen (Moore, A.), 87–88
Watergate scandal, 61, 103–104
Watts, Duncan, 39, 45–48, 49, 59
Wigan, Gareth, 56
Wilkinson, David, 61
Williams, Billy Dee, 24
Williams, John, 4
The Witch of Lime Street (Jaher), 170
The Wizard of Oz, 34
Wood, Gordon, 136, 146
Wookies (characters), 14, 37
Wordsworth, William, 49–50

X

Xenos (character), 13–14
Xerxes (character), 14

Y

Yoda (character)
    on another, 23–24
    as Buddha, 85–86
    death of, 29
    on despair of Dark Side, 101
    on feeling rapture, 179
    on future, 6
    training by, 98
    as wizened, 175
    working on mind, 174

Z

Zaentz, Saul, 57